www.EffortlessMath.com

... So Much More Online!

- ✓ FREE Math lessons
- ✓ More Math learning books!
- ✓ Mathematics Worksheets
- ✓ Online Math Tutors

Need a PDF version of this book?

Send email to: Info@EffortlessMath.com

Prepare for the ACCUPLACER Next Generation Math Test in 7 Days

A Quick Study Guide with Two Full-Length ACCUPLACER Math Practice Tests

By

Reza Nazari & Ava Ross

Reza Nazari & Ava Ross

All inquiries should be addressed to:

info@effortlessMath.com

www.EffortlessMath.com

ISBN–13: 978-1-64612-127-4

ISBN–10: 1-64612-127-9

Published by: Effortless Math Education

www.EffortlessMath.com

Description

Prepare for the ACCUPLACER Next Generation Math Test in 7 Days, which reflects the 2019 and 2020 test guidelines and topics, incorporates the best method and the right strategies to help you hone your math skills, overcome your exam anxiety, and boost your confidence -- and do your best to defeat ACCUPLACER Math test quickly. This quick study guide contains only the most important and critical math concepts a student will need in order to succeed on the ACCUPLACER Math test. Math concepts in this book break down the topics, so the material can be quickly grasped. Examples are worked step–by–step to help you learn exactly what to do.

This ACCUPLACER Math new edition has been updated to duplicate questions appearing on the most recent ACCUPLACER Math tests. It contains easy–to–read essential summaries that highlight the key areas of the ACCUPLACER Math test. You only need to spend about 3 – 5 hours daily in your 7–day period in order to achieve your goal. After reviewing this book, you will have solid foundation and adequate practice that is necessary to fully prepare for the ACCUPLACER Math.

Prepare for the ACCUPLACER Next Generation Math Test in 7 Days is for all ACCUPLACER Math test takers. It is a breakthrough in Math learning — offering a winning formula and the most powerful methods for learning basic Math topics confidently. Each section offers step–by–step instruction and helpful hints, with a few topics being tackled each day.

Inside the pages of this comprehensive book, students can learn math topics in a structured manner with a complete study program to help them understand essential math skills. It also has many exciting features, including:

- Content 100% aligned with the 2019-2020 ACCUPLACER test
- Written by ACCUPLACER Math tutors and test experts
- Complete coverage of all ACCUPLACER Math concepts and topics which you will be tested
- Step-by-step guide for all ACCUPLACER Math topics
- Dynamic design and easy-to-follow activities
- Over 1,500 additional ACCUPLACER math practice questions in both multiple-choice and grid-in formats with answers grouped by topic, so you can focus on your weak areas
- 2 full-length practice tests (featuring new question types) with detailed answers

Effortlessly and confidently follow the step–by–step instructions in this book to prepare for the ACCUPLACER Math in a short period of time.

Prepare for the ACCUPLACER Next Generation Math Test in 7 Days is the only book you'll ever need to master Basic Math topics! It can be used as a self–study course – you do not need to work with a Math tutor. (It can also be used with a Math tutor).

Ideal for self–study as well as for classroom usage.

About the Author

Reza Nazari is the author of more than 100 Math learning books including:
– **Math and Critical Thinking Challenges:** For the Middle and High School Student
– **GRE Math in 30 Days**
– **ASVAB Math Workbook 2018 - 2019**
– **Effortless Math Education Workbooks**
– **and many more Mathematics books ...**

Reza is also an experienced Math instructor and a test–prep expert who has been tutoring students since 2008. Reza is the founder of Effortless Math Education, a tutoring company that has helped many students raise their standardized test scores—and attend the colleges of their dreams. Reza provides an individualized custom learning plan and the personalized attention that makes a difference in how students view math.

You can contact Reza via email at:
reza@EffortlessMath.com

Find Reza's professional profile at:
goo.gl/zoC9rJ

Contents

Day 1: Integers, Ratios, and Proportions

Math Topics that you'll learn today:

- ✓ Adding and Subtracting Integers
- ✓ Multiplying and Dividing Integers
- ✓ Order of Operations
- ✓ Integers and Absolute Value
- ✓ Simplifying Ratios
- ✓ Proportional Ratios
- ✓ Similarity and Ratios

Without mathematics, there's nothing you can do. Everything around you is mathematics. Everything around you is numbers." - Shakuntala Devi

Adding and Subtracting Integers

Step-by-step guide:

- ✓ Integers includes: zero, counting numbers, and the negative of the counting numbers. $\{\ldots, -3, -2, -1, 0, 1, 2, 3, \ldots\}$
- ✓ Add a positive integer by moving to the right on the number line.
- ✓ Add a negative integer by moving to the left on the number line.
- ✓ Subtract an integer by adding its opposite.

Examples:

1) Solve. $(-8) - (-5) =$

   Keep the first number and convert the sign of the second number to its opposite. (change subtraction into addition. Then: $(-8) + 5 = -3$

2) Solve. $10 + (4 - 8) =$

   First subtract the numbers in brackets, $4 - 8 = -4$

   Then: $10 + (-4) =$ → change addition into subtraction: $10 - 4 = 6$

Multiplying and Dividing Integers

Step-by-step guide:

Use these rules for multiplying and dividing integers:

- ✓ (negative) × (negative) = positive
- ✓ (negative) ÷ (negative) = positive
- ✓ (negative) × (positive) = negative
- ✓ (negative) ÷ (positive) = negative
- ✓ (positive) × (positive) = positive

Examples:

1) Solve. $(2-5)\times(3)=$

First subtract the numbers in brackets, $2-5=-3 \rightarrow (-3)\times(3)=$

Now use this formula: (negative) × (positive) = negative
$(-3)\times(3)=-9$

2) Solve. $(-12)+(48\div 6)=$

First divided 48 by 6 , the numbers in brackets, $48\div 6=8$

$=(-12)+(8)=-12+8=-4$

Order of Operations

Step-by-step guide:

When there is more than one math operation, use PEMDAS:

- ✓ Parentheses
- ✓ Exponents
- ✓ Multiplication and Division (from left to right)
- ✓ Addition and Subtraction (from left to right)

Examples:

1) Solve. $(5+7)\div(3^2\div 3)=$

First simplify inside parentheses: $(12)\div(9\div 3)=(12)\div(3)=$
Then: $(12)\div(3)=4$

2) Solve. $(11\times 5)-(12-7)=$

First simplify inside parentheses: $(11\times 5)-(12-7)=(55)-(5)=$

Then: $(55)-(5)=50$

Integers and Absolute Value

Step-by-step guide:

- ✓ To find an absolute value of a number, just find its distance from 0 on number line! For example, the distance of 12 and -12 from zero on number line is 12!

Examples:

1) Solve. $\frac{|-18|}{9} \times |5-8| =$

First find $|-18|$, →the absolute value of -18 is 18, then: $|-18| = 18$

$\frac{18}{9} \times |5-8| =$

Next, solve $|5-8|$, → $|5-8| = |-3|$, the absolute value of -3 is 3. $|-3| = 3$

Then: $\frac{18}{9} \times 3 = 2 \times 3 = 6$

2) Solve. $|10-5| \times \frac{|-2\times 6|}{3} =$

First solve $|10-5|$, →$|10-5| = |5|$, the absolute value of 5 is 5, $|5| = 5$

$5 \times \frac{|-2\times 6|}{3} =$

Now solve $|-2 \times 6|$, → $|-2 \times 6| = |-12|$, the absolute value of -12 is 12, $|-12| = 12$

Then: $5 \times \frac{12}{3} = 5 \times 4 = 20$

Simplifying Ratios

Step-by-step guide:

- ✓ Ratios are used to make comparisons between two numbers.
- ✓ Ratios can be written as a fraction, using the word "to", or with a colon.
- ✓ You can calculate equivalent ratios by multiplying or dividing both sides of the ratio by the same number.

Examples:

1) Simplify. $8:4 =$

Both numbers 8 and 4 are divisible by 4, $\Rightarrow 8 \div 4 = 2, 4 \div 4 = 1,$

Then: $8:4 = 2:1$

2) Simplify. $\frac{12}{36} =$

Both numbers 12 and 36 are divisible by 12, $\Rightarrow$ $12 \div 12 = 1, 36 \div 12 = 3,$

Then: $\frac{12}{36} = \frac{1}{3}$

Proportional Ratios

Step-by-step guide:

- ✓ A proportion means that two ratios are equal. It can be written in two ways: $\frac{a}{b} = \frac{c}{d}$, $a : b = c : d$
- ✓ The proportion $\frac{a}{b} = \frac{c}{d}$ can be written as: $a \times d = c \times b$

Examples:

1) Solve this proportion for x. $\frac{4}{8} = \frac{5}{x}$

   Use cross multiplication: $\frac{4}{8} = \frac{5}{x} \Rightarrow 4 \times x = 5 \times 8 \Rightarrow 4x = 40$

   Divide to find x: $x = \frac{40}{4} \Rightarrow x = 10$

2) If a box contains red and blue balls in ratio of 2: 3 red to blue, how many red balls are there if 90 blue balls are in the box?

   Write a proportion and solve. $\frac{2}{3} = \frac{x}{90}$

   Use cross multiplication: $2\times 90 = 3\times x \Rightarrow 180 = 3x$

   Divide to find x: $x = \frac{180}{3} \Rightarrow x = 60$

Similarity and Ratios

Step-by-step guide:

- ✓ Two or more figures are similar if the corresponding angles are equal, and the corresponding sides are in proportion.

Examples:

1) A girl $160\ cm$ tall, stands $360\ cm$ from a lamp post at night. Her shadow from the light is $90\ cm$ long. How high is the lamp post?

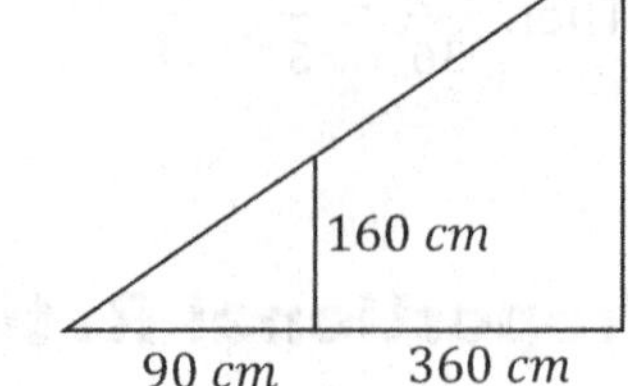

Write the proportion and solve for missing side.

$$\frac{\text{Smaller triangle height}}{\text{Smaller triangle base}} = \frac{\text{Bigger triangle height}}{\text{Bigger triangle base}}$$

$$\Rightarrow \frac{90cm}{160cm} = \frac{90+360cm}{x} \Rightarrow 90x = 160 \times 450 \Rightarrow x = 800\ cm$$

2) A tree $32\ feet$ tall casts a shadow $12\ feet$ long. Jack is $6\ feet$ tall. How long is Jack's shadow?

Write a proportion and solve for the missing number.

$$\frac{32}{12} = \frac{6}{x} \Rightarrow 32x = 6 \times 12 = 72 \Rightarrow 32x = 72 \rightarrow x = \frac{72}{32} = 2.25\ feet$$

Day 1 Practices

Find each sum or difference.

1) $15 + (-8) =$

2) $(-11) + (-21) =$

3) $7 + (-27) =$

4) $45 + (-14) =$

5) $(-8) + (-12) + 6 =$

6) $37 + (-16) + 12 =$

Find each product or quotient.

7) $(-7) \times (-8) =$

8) $4 \times (-5) =$

9) $5 \times (-11) =$

10) $(-5) \times (-20) =$

11) $-(2) \times (-8) \times 3 =$

12) $(12 - 4) \times (-10) =$

Evaluate each expression.

13) $5 + (6 \times 3) =$

14) $13 - (2 \times 5) =$

15) $(14 \times 2) + 18 =$

16) $(12 - 5) - (4 \times 3) =$

17) $25 + (14 \div 2) =$

18) $(18 \times 5) \div 2 =$

Evaluate the value.

19) $8 - |4 - 18| - |-2| =$

20) $|-2| - \frac{|-20|}{4} =$

21) $\frac{|-66|}{11} \times |-6| =$

22) $\frac{|-5 \times -3|}{5} \times \frac{|-20|}{4} =$

23) $|2 \times -4| + \frac{|-40|}{5} =$

24) $\frac{|-28|}{4} \times \frac{|-55|}{11} =$

Reduce each ratio.

25) $24:16 =$ ___:___

26) $4:40 =$ ___:___

27) $6:72 =$ ___:___

28) $18:36 =$ ___:___

29) $6:100 =$ ___:___

30) $4:24 =$ ___:___

Solve each proportion.

31) $\frac{4}{10} = \frac{14}{x}, x =$ ____

32) $\frac{2}{12} = \frac{7}{x}, x =$ ____

33) $\frac{3}{5} = \frac{27}{x}, x =$ ____

34) $\frac{1}{5} = \frac{x}{80}, x =$ ____

35) $\frac{3}{7} = \frac{x}{63}, x =$ ____

36) $\frac{2}{8} = \frac{13}{x}, x =$ ____

Solve each problem.

37) Two rectangles are similar. The first is $6\ feet$ wide and $20\ feet$ long. The second is $15\ feet$ wide. What is the length of the second rectangle? ________________

38) Two rectangles are similar. One is $2.5\ meters$ by $9\ meters$. The longer side of the second rectangle is $22.5\ meters$. What is the other side of the second rectangle?

____

Answers

1) 7
2) −32
3) −20
4) 31
5) −14
6) 33
7) 56
8) −20
9) −55
10) 100
11) 48
12) −80
13) 23
14) 3
15) 46
16) −5
17) 32
18) 45
19) −8
20) −3
21) 36
22) 15
23) 16
24) 35
25) 3: 2
26) 1: 10
27) 1: 12
28) 1: 2
29) 3: 50
30) 1: 6
31) 35
32) 42
33) 45
34) 16
35) 27
36) 52
37) 50 feet
38) 6.25 meters

Day 2: Percentage, Exponents, Variables and Roots

Math Topics that you'll learn today:

- ✓ Percentage Calculations
- ✓ Percent Problems
- ✓ Percent of Increase and Decrease
- ✓ Simple Interest
- ✓ Multiplication Property of Exponents
- ✓ Division Property of Exponents
- ✓ Powers of Products and Quotients
- ✓ Zero and Negative Exponents
- ✓ Negative Exponents and Negative Bases
- ✓ Scientific Notation
- ✓ Square Roots

Mathematics is no more computation than typing is literature. – John Allen Paulos

Percentage Calculations

Step-by-step guide:

- ✓ Percent is a ratio of a number and 100. It always has the same denominator, 100. Percent symbol is %.
- ✓ Percent is another way to write decimals or fractions. For example:

$$40\% = 0.40 = \frac{40}{100} = \frac{2}{5}$$

- ✓ Use the following formula to find part, whole, or percent:

$$\text{part} = \frac{\text{percent}}{100} \times \text{whole}$$

Examples:

1) What is 10% of 45? Use the following formula: $\text{part} = \frac{\text{percent}}{100} \times \text{whole}$

$\text{part} = \frac{10}{100} \times 45 \rightarrow \text{part} = \frac{1}{10} \times 45 \rightarrow \text{part} = \frac{45}{10} \rightarrow \text{part} = 4.5$

2) What is 15% of 24? Use the percent formula: $\text{part} = \frac{\text{percent}}{100} \times \text{whole}$

$\text{part} = \frac{15}{100} \times 24 \rightarrow \text{part} = \frac{360}{100} \rightarrow \text{part} = 3.6$

Percent Problems

Step-by-step guide:

- ✓ In each percent problem, we are looking for the base, or part or the percent.
- ✓ Use the following equations to find each missing section.
  - Base = Part ÷ Percent
  - Part = Percent × Base
  - Percent = Part ÷ Base

Examples:

1) 1.2 is what percent of 24?

In this problem, we are looking for the percent. Use the following equation:

$$Percent = Part \div Base \rightarrow Percent = 1.2 \div 24 = 0.05 = 5\%$$

2) 20 is 5% of what number?

Use the following formula: $Base = Part \div Percent \rightarrow Base = 20 \div 0.05 = 400$
20 is 5% of 400.

Percent of Increase and Decrease

Step-by-step guide:

To find the percentage of increase or decrease:
- ✓ New Number - Original Number
- ✓ The result ÷ Original Number × 100
- ✓ If your answer is a negative number, then this is a percentage decrease. If it is positive, then this is a percent of increase.

Examples:

1) Increased by 50%, the numbers 84 becomes:

First find 50% of 84 $\rightarrow$ $\frac{50}{100} \times 84 = \frac{50 \times 84}{100} = 42$

Then: $84 + 42 = 126$

2) The price of a shirt increases from $\$10$ to $\$14$. What is the percent increase?

First: $14 - 10 = 4$

4 is the result. Then: $4 \div 10 = \frac{4}{10} = 0.4 = 40\%$

Simple Interest

Step-by-step guide:

- ✓ Simple Interest: The charge for borrowing money or the return for lending it.
  To solve a simple interest problem, use this formula:
  Interest = principal x rate x time $\Rightarrow$ $I = p \times r \times t$

Examples:

1) Find simple interest for $\$450$ investment at 7% for 8 years.

Use Interest formula: $I = prt$, $P = \$450$, $r = 7\% = \frac{7}{100} = 0.07$ and $t = 8$

Then: $I = 450 \times 0.07 \times 8 = \$252$

2) Find simple interest for $\$5{,}200$ at 4% for 3 years.

Use Interest formula: $I = prt$, $P = \$5,200$, $r = 4\% = \frac{4}{100} = 0.04$ and $t = 3$
Then: $I = 5,200 \times 0.04 \times 3 = \$624$

Multiplication Property of Exponents

Step-by-step guide:

- ✓ Exponents are shorthand for repeated multiplication of the same number by itself. For example, instead of 2×2, we can write 2^2. For $3 \times 3 \times 3 \times 3$, we can write 3^4
- ✓ In algebra, a variable is a letter used to stand for a number. The most common letters are: $x, y, z, a, b, c, m, and\ n$.
- ✓ Exponent's rules: $x^a \times x^b = x^{a+b}$, $\frac{x^a}{x^b} = x^{a-b}$

$$(x^a)^b = x^{a \times b}, \qquad (xy)^a = x^a \times y^a \text{ , } (\frac{a}{b})^c = \frac{a^c}{b^c}$$

Examples:

1) Multiply. $-2x^5 \times 7x^3 =$
Use Exponent's rules: $x^a \times x^b = x^{a+b} \rightarrow x^5 \times x^3 = x^{5+3} = x^8$
Then: $-2x^5 \times 7x^3 = -14x^8$

2) Multiply. $(x^2y^4)^3 =$
Use Exponent's rules: $(x^a)^b = x^{a \times b}$. Then: $(x^2\ y^4)^3 = x^{2\times3}y^{4\times3} = x^6y^{12}$

Division Property of Exponents

Step-by-step guide:

- ✓ For division of exponents use these formulas: $\frac{x^a}{x^b} = x^{a-b}$, $x \neq 0$

$$\frac{x^a}{x^b} = \frac{1}{x^{b-a}}, x \neq 0, \qquad \frac{1}{x^b} = x^{-b}$$

Examples:

1) Simplify. $\frac{4x^3y}{36x^2y^3} =$

First cancel the common factor: $4 \rightarrow \frac{4x^3y}{36x^2y^3} = \frac{x^3y}{9x^2y^3}$

Use Exponent's rules: $\frac{x^a}{x^b} = x^{a-b} \rightarrow \frac{x^3}{x^2} = x^{3-2}$

Then: $\frac{4x^3y}{36x^2y^3} = \frac{xy}{9y^3} \rightarrow$ now cancel the common factor: $y \rightarrow \frac{xy}{9y^3} = \frac{x}{9y^2}$

2) Divide. $\frac{2x^{-5}}{9x^{-2}} =$

Use Exponent's rules: $\frac{x^a}{x^b} = \frac{1}{x^{b-a}} \rightarrow \frac{x^{-5}}{x^{-2}} = \frac{1}{x^{-2-(-5)}} = \frac{1}{x^{-2+5}} = \frac{1}{x^3}$

Then: $\frac{2x^{-5}}{9x^{-2}} = \frac{2}{9x^3}$

Powers of Products and Quotients

Step-by-step guide:

✓ For any nonzero numbers a and b and any integer x, $(ab)^x = a^x \times b^x$.

Examples:

1) Simplify. $(3x^5y^4)^2 =$

Use Exponent's rules: $(x^a)^b = x^{a\times b}$

$(3x^5y^4)^2 = (3)^2(x^5)^2(y^4)^2 = 9x^{5\times2}y^{4\times2} = 9x^{10}y^8$

2) Simplify. $(\frac{2x}{3x^2})^2 =$ First cancel the common factor: $x \rightarrow \left(\frac{2x}{3x^2}\right)^2 = \left(\frac{2}{3x}\right)^2$

Use Exponent's rules: $(\frac{a}{b})^c = \frac{a^c}{b^c}$, Then: $(\frac{2}{3x})^2 = \frac{2^2}{(3x)^2} = \frac{4}{9x^2}$

Zero and Negative Exponents

Step-by-step guide:

✓ A negative exponent simply means that the base is on the wrong side of the fraction line, so you need to flip the base to the other side. For instance, "x^{-2}" (pronounced as "ecks to the minus two") just means "x^2" but underneath, as in $\frac{1}{x^2}$.

Examples:

1) Evaluate. $(\frac{4}{9})^{-2} =$

Use Exponent's rules: $\frac{1}{x^b} = x^{-b} \rightarrow (\frac{4}{9})^{-2} = \frac{1}{(\frac{4}{9})^2} = \frac{1}{\frac{4^2}{9^2}}$

Now use fraction rule: $\frac{1}{\frac{b}{c}} = \frac{c}{b} \rightarrow \frac{1}{\frac{4^2}{9^2}} = \frac{9^2}{4^2} = \frac{81}{16}$

2) Evaluate. $(\frac{5}{6})^{-3} =$

Use Exponent's rules: $\frac{1}{x^b} = x^{-b} \rightarrow (\frac{5}{6})^{-3} = \frac{1}{(\frac{5}{6})^3} = \frac{1}{\frac{5^3}{6^3}}$

Now use fraction rule: $\frac{1}{\frac{b}{c}} = \frac{c}{b} \rightarrow \frac{1}{\frac{5^3}{6^3}} = \frac{6^3}{5^3} = \frac{216}{125}$

Negative Exponents and Negative Bases

Step-by-step guide:

✓ Make the power positive. A negative exponent is the reciprocal of that number with a positive exponent.

✓ The parenthesis is important!

✓ 5^{-2} is not the same as $(-5)^{-2}$

$$(-5)^{-2} = -\frac{1}{5^2} \text{ and } (-5)^{-2} = +\frac{1}{5^2}$$

Examples:

1) Simplify. $(\frac{3a}{2c})^{-2} =$

Use Exponent's rules: $\frac{1}{x^b} = x^{-b} \rightarrow (\frac{3a}{2c})^{-2} = \frac{1}{(\frac{3a}{2c})^2} = \frac{1}{\frac{3^2a^2}{2^2c^2}}$

Now use fraction rule: $\frac{1}{\frac{b}{c}} = \frac{c}{b} \rightarrow \frac{1}{\frac{3^2a^2}{2^2c^2}} = \frac{2^2c^2}{3^2a^2}$

Then: $\frac{2^2c^2}{3^2a^2} = \frac{4c^2}{9a^2}$

2) Simplify. $(-\frac{5x}{3yz})^{-3} =$

Use Exponent's rules: $\frac{1}{x^b} = x^{-b} \rightarrow (-\frac{5x}{3yz})^{-3} = \frac{1}{(-\frac{5x}{3yz})^3} = \frac{1}{-\frac{5^3x^3}{3^3y^3z^3}}$

Now use fraction rule: $\frac{1}{\frac{b}{c}} = \frac{c}{b} \rightarrow \frac{1}{-\frac{5^3x^3}{3^3y^3z^3}} = -\frac{3^3y^3z^3}{5^3x^3} = -\frac{27y^3z^3}{125x^3}$

Scientific Notation

Step-by-step guide:

- ✓ It is used to write very big or very small numbers in decimal form.
- ✓ In scientific notation all numbers are written in the form of:

$$m \times 10^n$$

Decimal notation	Scientific notation
5	5×10^0
$-25{,}000$	-2.5×10^4
0.5	5×10^{-1}
2,122.456	2.122456×10^3

Examples:

1) Write $\mathbf{0.00012}$ in scientific notation.

First, move the decimal point to the right so that you have a number that is between 1 and 10. Then: $N = 1.2$

Second, determine how many places the decimal moved in step 1 by the power of 10.

Then: 10^{-4} → When the decimal moved to the right, the exponent is negative.

Then: $0.00012 = 1.2 \times 10^{-4}$

2) Write $\mathbf{8.3 \times 10^{-5}}$ in standard notation.

10^{-5} → When the decimal moved to the right, the exponent is negative.

Then: $8.3 \times 10^{-5} = 0.000083$

Square Roots

Step-by-step guide:

- ✓ A square root of x is a number r whose square is: $r^2 = x$

  r is a square root of x.

Examples:

1) Find the square root of $\sqrt{225}$.

   First factor the number: $225 = 15^2$, Then: $\sqrt{225} = \sqrt{15^2}$

   Now use radical rule: $\sqrt[n]{a^n} = a$

   Then: $\sqrt{15^2} = 15$

2) Evaluate. $\sqrt{4} \times \sqrt{16} =$

   First factor the numbers: $4 = 2^2$ and $16 = 4^2$

   Then: $\sqrt{4} \times \sqrt{16} = \sqrt{2^2} \times \sqrt{4^2}$

   Now use radical rule: $\sqrt[n]{a^n} = a$, Then: $\sqrt{2^2} \times \sqrt{4^2} = 2 \times 4 = 8$

Day 2 Practices

Calculate the given percent of each value.

1) 5% of 60 = ____
2) 10% of 30 = ____
3) 20% of 25 = ____
4) 50% of 80 = ____
5) 40% of 200 = ____
6) 20% of 45 = ____

Solve each problem.

7) 20 is what percent of 50? ____%
8) 18 is what percent of 90? ____%
9) 12 is what percent of 15? ____%
10) 16 is what percent of 200? ____%
11) 24 is what percent of 800? ____%
12) 48 is what percent of 400? ____%

Solve each percent of change word problem.

13) Bob got a raise, and his hourly wage increased from $12 to $15. What is the percent increase? ______ %

14) The price of a pair of shoes increases from $20 to $32. What is the percent increase? ____ %

Determine the simple interest for these loans.

15) $1,300 at 5% for 6 years. $ _______
16) $5,400 at 3.5% for 6 months. $ _______

Simplify and write the answer in exponential form.

17) $2yx^3 \times 4x^2y^3 =$
18) $4x^2 \times 9x^3y^4 =$
19) $7x^4y^5 \times 3x^2y^3 =$
20) $9x^2y^5 \times 7xy^3 =$
21) $4xy^4 \times 7x^3y^3 =$
22) $8x^2y^3 \times 3x^5y^3 =$

Simplify. (Division Property of Exponents)

23) $\frac{3^7 \times 3^4}{3^8 \times 3^2} =$
24) $\frac{5x}{10x^3} =$
25) $\frac{6x^3}{4x^5} =$
26) $\frac{24x^3}{28x^6} =$
27) $\frac{24x^3}{18y^8} =$
28) $\frac{50xy^4}{10y^2} =$

Simplify. (Powers of Products and Quotients)

29) $(9x^7y^5)^2 =$

30) $(4x^4y^6)^5 =$

31) $(3x \times 4y^3)^2 =$

32) $(\frac{5x}{x^2})^2 =$

33) $\left(\frac{x^4y^4}{x^2y^2}\right)^3 =$

34) $\left(\frac{25x}{5x^6}\right)^2 =$

Evaluate the following expressions. (Zero and Negative Exponents)

35) $2^{-3} =$

36) $3^{-3} =$

37) $7^{-3} =$

38) $6^{-3} =$

39) $8^{-3} =$

40) $9^{-2} =$

Simplify. (Negative Exponents and Negative Bases)

41) $-5x^{-2}y^{-3} =$

42) $20x^{-4}y^{-1} =$

43) $14a^{-6}b^{-7} =$

44) $-12x^2y^{-3} =$

45) $-\frac{25}{x^{-6}} =$

46) $\frac{7b}{-9c^{-4}} =$

Write each number in scientific notation.

47) $0.000325 =$

48) $0.00023 =$

49) $56{,}000{,}000 =$

50) $21{,}000 =$

Evaluate.

51) $\sqrt{9} \times \sqrt{4} =$ ________

52) $\sqrt{64} \times \sqrt{25} =$ ________

53) $\sqrt{8} \times \sqrt{2} =$ ________

54) $\sqrt{6} \times \sqrt{6} =$ ________

55) $\sqrt{5} \times \sqrt{5} =$ ________

56) $\sqrt{8} \times \sqrt{8} =$ ________

Answers

1) 3
2) 3
3) 5
4) 40
5) 80
6) 9
7) 40%
8) 20%
9) 80%
10) 8%
11) 3%
12) 12%
13) 25%
14) 60%
15) $390
16) $94.50
17) $8x^5y^4$
18) $36x^5y^4$
19) $21x^6y^8$
20) $63x^3y^8$
21) $28x^4y^7$
22) $24x^7y^6$
23) 3
24) $\frac{1}{2x^2}$
25) $\frac{3}{2x^2}$
26) $\frac{6}{7x^3}$
27) $\frac{4x^3}{3y^8}$
28) $5xy^2$
29) $81x^{14}y^{10}$
30) $1,024x^{20}y^{30}$
31) $144x^2y^6$
32) $\frac{25}{x^2}$
33) x^6y^6
34) $\frac{25}{x^{10}}$
35) $\frac{1}{8}$
36) $\frac{1}{27}$
37) $\frac{1}{343}$
38) $\frac{1}{216}$
39) $\frac{1}{512}$
40) $\frac{1}{81}$
41) $-\frac{5}{x^2y^3}$
42) $\frac{20}{x^4y}$
43) $\frac{14}{a^6b^7}$
44) $-\frac{12x^2}{y^3}$
45) $-25x^6$
46) $-\frac{7bc^4}{9}$
47) 3.25×10^{-4}
48) 2.3×10^{-4}
49) 5.6×10^7
50) 2.1×10^4
51) 6
52) 40
53) 4
54) 6
55) 5
56) 8

Day 3: Expressions, Variables, Equations and Inequalities

Math Topics that you'll learn today:

- ✓ Simplifying Variable Expressions
- ✓ Simplifying Polynomial Expressions
- ✓ The Distributive Property
- ✓ Evaluating One Variable
- ✓ Evaluating Two Variables
- ✓ Combining like Terms
- ✓ One–Step Equations
- ✓ Multi–Step Equations
- ✓ Graphing Single–Variable Inequalities
- ✓ One–Step Inequalities
- ✓ Multi–Step Inequalities

Mathematics is, as it were, a sensuous logic, and relates to philosophy as do the arts, music, and plastic art to poetry. ~ K. Shegel

Simplifying Variable Expressions

Step-by-step guide:

- ✓ In algebra, a variable is a letter used to stand for a number. The most common letters are: $x, y, z, a, b, c, m, and\ n$.
- ✓ algebraic expression is an expression contains integers, variables, and the math operations such as addition, subtraction, multiplication, division, etc.
- ✓ In an expression, we can combine "like" terms. (values with same variable and same power)

Examples:

1) Simplify this expression. $(10x + 2x + 3) = ?$
   Combine like terms. Then: $(10x + 2x + 3) = 12x + 3$ (remember you cannot combine variables and numbers.
2) Simplify this expression. $12 - 3x^2 + 9x + 5x^2 = ?$
   Combine "like" terms: $-3x^2 + 5x^2 = 2x^2$
   Then: $12 - 3x^2 + 9x + 5x^2 = 12 + 2x^2 + 9x$. Write in standard form (biggest powers first): $2x^2 + 9x + 12$

Simplifying Polynomial Expressions

Step-by-step guide:

- ✓ In mathematics, a polynomial is an expression consisting of variables and coefficients that involves only the operations of addition, subtraction, multiplication, and non-negative integer exponents of variables.

$$P(x) = a_n x^n + a_{n-1}x^{n-1} + \ldots + a_2x^2 + a_1x + a_0$$

Examples:

1) Simplify this Polynomial Expressions. $4x^2 - 5x^3 + 15x^4 - 12x^3 =$
   Combine "like" terms: $-5x^3 - 12x^3 = -17x^3$
   Then: $4x^2 - 5x^3 + 15x^4 - 12x^3 = 4x^2 - 17x^3 + 15x^4$
   Then write in standard form: $4x^2 - 17x^3 + 15x^4 = 15x^4 - 17x^3 + 4x^2$
2) Simplify this expression. $(2x^2 - x^4) - (4x^4 - x^2) =$
   First use distributive property: $\rightarrow$ multiply $(-)$ into $(4x^4 - x^2)$

$(2x^2 - x^4) - (4x^4 - x^2) = 2x^2 - x^4 - 4x^4 + x^2$
Then combine "like" terms: $2x^2 - x^4 - 4x^4 + x^2 = 3x^2 - 5x^4$
And write in standard form: $3x^2 - 5x^4 = -5x^4 + 3x^2$

The Distributive Property

Step-by-step guide:

✓ Distributive Property:

$$a(b + c) = ab + ac$$

Examples:

1) Simply. $(5x - 3)(-5) =$

Use Distributive Property formula: $a(b + c) = ab + ac$
$(5x - 3)(-5) = -25x + 15$

2) Simply$(-8)(2x - 8) =$

Use Distributive Property formula: $a(b + c) = ab + ac$
$(-8)(2x - 8) = -16x + 64$

Evaluating One Variable

Step-by-step guide:

✓ To evaluate one variable expression, find the variable and substitute a number for that variable.

✓ Perform the arithmetic operations.

Examples:

1) Solve this expression. $12 - 2x\ , x = -1$

First substitute -1 for x, then:

$12 - 2x = 12 - 2(-1) = 12 + 2 = 14$

2) Solve this expression. $-8 + 5x\ , x = 3$

First substitute 3 for x, then:

$-8 + 5x = -8 + 5(3) = -8 + 15 = 7$

Evaluating Two Variables

Step-by-step guide:

- ✓ To evaluate an algebraic expression, substitute a number for each variable and perform the arithmetic operations.

Examples:

1) Solve this expression. $-3x + 5y, x = 2, y = -1$

First substitute 2 for x, and -1 for y, then:

$-3x + 5y = -3(2) + 5(-1) = -6 - 5 = -11$

2) Solve this expression. $2(a - 2b), a = -1, b = 3$

First substitute -1 for a, and 3 for b, then:

$2(a - 2b) = 2a - 4b = 2(-1) - 4(3) = -2 - 12 = -14$

Combining like Terms

Step-by-step guide:

- ✓ Terms are separated by "+" and "-" signs.
- ✓ Like terms are terms with same variables and same powers.
- ✓ Be sure to use the "+" or "-" that is in front of the coefficient.

Examples:

1) Simplify this expression. $(-5)(8x - 6) =$

Use Distributive Property formula: $a(b + c) = ab + ac$
$(-5)(8x - 6) = -40x + 30$

2) Simplify this expression. $(-3)(2x - 2) + 6 =$

First use Distributive Property formula: $a(b + c) = ab + ac$
$(-3)(2x - 2) + 6 = -6x + 6 + 6$

And Combining like Terms:

$-6x + 6 + 6 = -6x + 12$

One–Step Equations

Step-by-step guide:

- ✓ The values of two expressions on both sides of an equation are equal. $ax + b = c$
- ✓ You only need to perform one Math operation in order to solve the one-step equations.
- ✓ To solve one-step equation, find the inverse (opposite) operation is being performed.
- ✓ The inverse operations are:
  - Addition and subtraction
  - Multiplication and division

Examples:

1) Solve this equation. $x + 24 = 0\,, x = ?$
   Here, the operation is addition and its inverse operation is subtraction. To solve this equation, subtract 24 from both sides of the equation: $x + 24 - 24 = 0 - 24$
   Then simplify: $x + 24 - 24 = 0 - 24 \;\rightarrow\; x = -24$

2) Solve this equation. $3x = 15, x = ?$
   Here, the operation is multiplication (variable x is multiplied by 3) and its inverse operation is division. To solve this equation, divide both sides of equation by 3:

$$3x = 15 \rightarrow \frac{3x}{3} = \frac{15}{3} \rightarrow x = 5$$

Multi–Step Equations

Step-by-step guide:

- ✓ Combine "like" terms on one side.
- ✓ Bring variables to one side by adding or subtracting.
- ✓ Simplify using the inverse of addition or subtraction.
- ✓ Simplify further by using the inverse of multiplication or division.

Examples:

1) Solve this equation. $-(2 - x) = 5$

First use Distributive Property: $-(2-x)=-2+x$
Now solve by adding 2 to both sides of the equation. $-2+x=5 \rightarrow -2+x+2=5+2$
Now simplify: $-2+x+2=5+2 \rightarrow x=7$

2) Solve this equation. $4x+10=25-x$

First bring variables to one side by adding x to both sides.
$4x+10+x=25-x+x \rightarrow 5x+10=25$. Now, subtract 10 from both sides:
$5x+10-10=25-10 \rightarrow 5x=15$
Now, divide both sides by 5: $5x=15 \rightarrow \frac{5x}{5}=\frac{15}{5} \rightarrow x=3$

Graphing Single–Variable Inequalities

Step-by-step guide:

- ✓ Inequality is similar to equations and uses symbols for "less than" (<) and "greater than" (>).
- ✓ To solve inequalities, we need to isolate the variable. (like in equations)
- ✓ To graph an inequality, find the value of the inequality on the number line.
- ✓ For less than or greater than draw open circle on the value of the variable.
- ✓ If there is an equal sign too, then use filled circle.
- ✓ Draw a line to the right or to the left for greater or less than.

Examples:

1) Draw a graph for $x>2$

-4 -2 0 2 4

Since, the variable is greater than 2, then we need to find 2 and draw an open circle above it. Then, draw a line to the right.

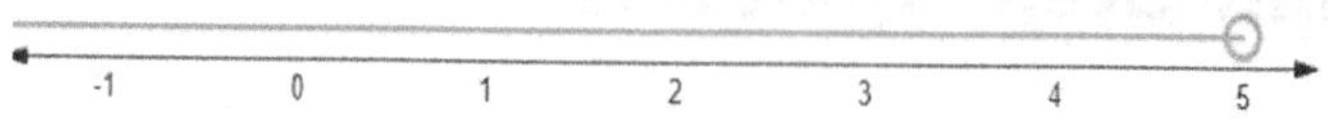

-1 0 1 2 3 4 5

2) Graph this inequality. $x<5$

One–Step Inequalities

Step-by-step guide:

- ✓ Similar to equations, first isolate the variable by using inverse operation.
- ✓ For dividing or multiplying both sides by negative numbers, flip the direction

of the inequality sign.

Examples:

Multi–Step Inequalities

1) Solve and graph the inequality. $x + 2 \geq 3$.

Subtract 2 from both sides. $x + 2 \geq 3 \rightarrow x + 2 - 2 \geq 3 - 2$, then: $x \geq 1$

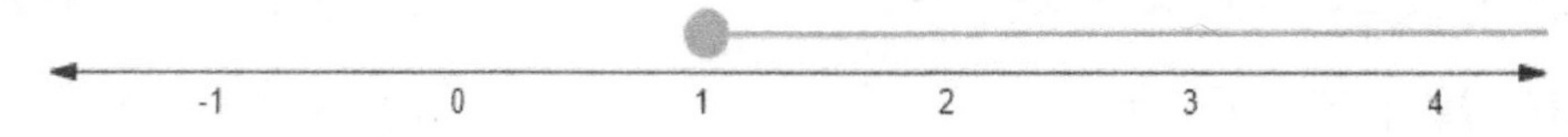

2) Solve this inequality. $x - 1 \leq 2$

Add 1 to both sides. $x - 1 \leq 2 \rightarrow x - 1 + 1 \leq 2 + 1$, then: $x \leq 3$

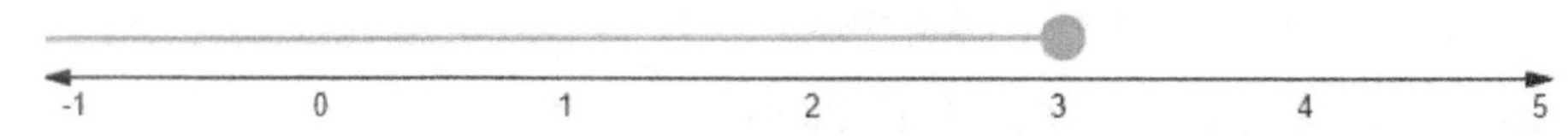

Step-by-step guide:

- ✓ Isolate the variable.
- ✓ Simplify using the inverse of addition or subtraction.
- ✓ Simplify further by using the inverse of multiplication or division.

Examples:

1) Solve this inequality. $2x - 2 \leq 6$

First add 2 to both sides: $2x - 2 + 2 \leq 6 + 2 \rightarrow 2x \leq 8$

Now, divide both sides by 2: $2x \leq 8 \rightarrow x \leq 4$

2) Solve this inequality. $2x - 4 \leq 8$

First add 4 to both sides: $2x - 4 + 4 \leq 8 + 4$

Then simplify: $2x - 4 + 4 \leq 8 + 4 \rightarrow 2x \leq 12$

Now divide both sides by 2: $\frac{2x}{2} \leq \frac{12}{2} \rightarrow x \leq 6$

Day 3 Practices

Simplify each expression.

1) $(2x + x + 8 + 19) =$

2) $(-22x - 26x + 24) =$

3) $8x + 3 - 4x =$

4) $-2 - 5x^2 - 2x^2 =$

5) $3 + 10x^2 + 2 =$

6) $3x^2 + 6x + 12x^2 =$

Simplify each polynomial.

7) $(2x^3 + 5x^2) - (12x + 2x^2) =$ ______________________

8) $(2x^5 + 2x^3) - (7x^3 + 6x^2) =$ ______________________

9) $(12x^4 + 4x^2) - (2x^2 - 6x^4) =$ ______________________

Use the distributive property to simply each expression.

10) $2(2 + 3x) =$

11) $3(5 + 5x) =$

12) $4(3x - 8) =$

13) $(6x - 2)(-2) =$

14) $(-3)(x + 2) =$

15) $(2 + 2x)5 =$

Evaluate each expression using the value given.

16) $5 + x, x = 2$

17) $x - 2, x = 4$

18) $8x + 1, x = 9$

19) $x - 12, x = -1$

20) $9 - x, x = 3$

21) $x + 2, x = 5$

Evaluate each expression using the values given.

22) $2x + 4y,\ x = 3, y = 2$

23) $8x + 5y,\ x = 1, y = 5$

24) $-2a + 4b,\ a = 6, b = 3$

25) $4x + 7 - 2y,\ x = 7, y = 6$

Simplify each expression. (Combining like Terms)

26) $2x + x + 2 =$

27) $2(5x - 3) =$

28) $7x - 2x + 8 =$

29) $(-4)(3x - 5) =$

30) $9x - 7x - 5 =$

31) $16x - 5 + 8x =$

Solve each equation. (One–Step Equations)

32) $16 = -4 + x, x =$ ____

33) $x - 4 = -25, x =$ ____

34) $x + 12 = -9, x =$ ____

35) $14 = 18 - x, x =$ ____

36) $2 + x = -14, x =$ ____

37) $x - 5 = 15, x =$ ____

Solve each equation. (Multi–Step Equations)

38) $-3(2 + x) = 3$

39) $-2(4 + x) = 4$

40) $20 = -(x - 8)$

41) $2(2 - 2x) = 20$

42) $-12 = -(2x + 8)$

43) $5(2 + x) = 5$

Draw a graph for each inequality.

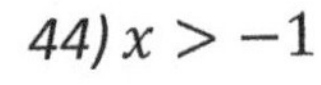
44) $x > -1$

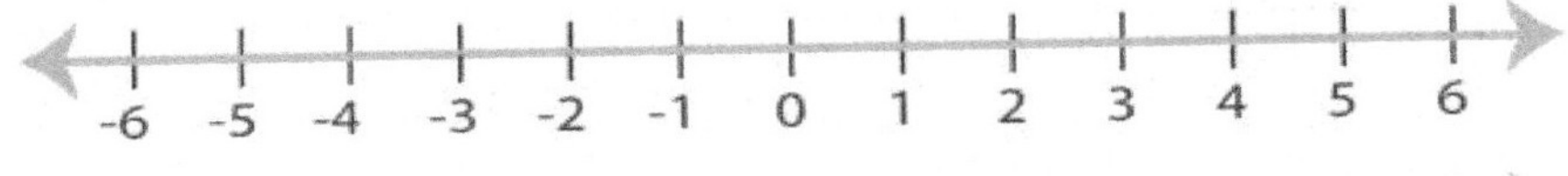

45) $x < 3$

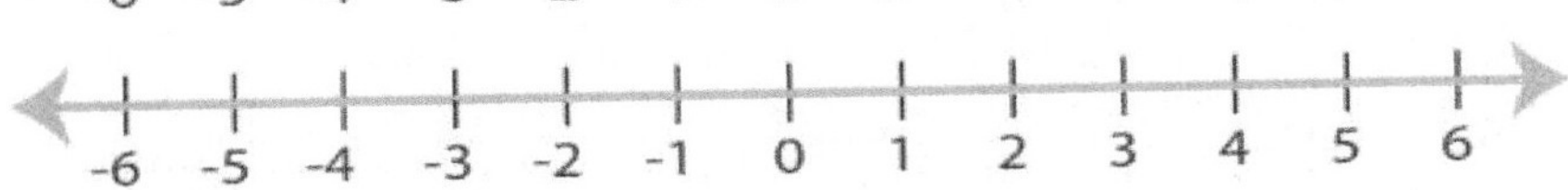

Solve each inequality and graph it.

46) $2x \geq 12$

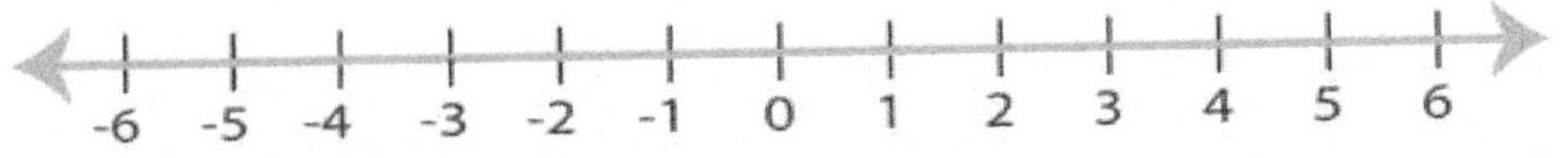

47) $4 + x \leq 5$

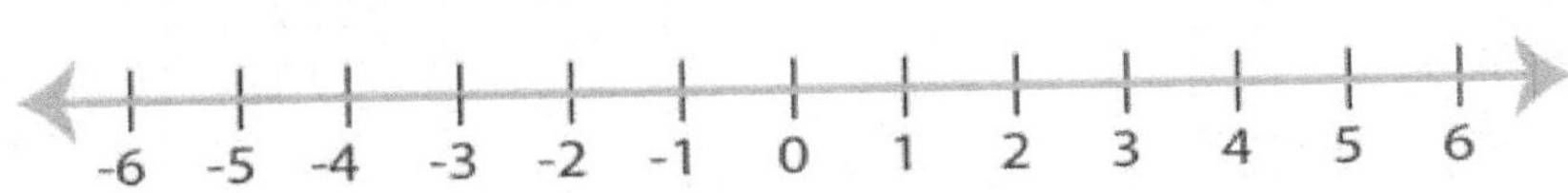

Solve each inequality.

48) $4x - 16 \leq 12$

49) $16x - 4 \leq 28$

50) $-15 + 9x \leq 30$

51) $2(x - 3) \leq 6$

52) $14x - 10 \leq 18$

53) $8x - 42 < 38$

Answers

1) $3x + 27$
2) $-48x + 24$
3) $4x + 3$
10) $6x + 4$
11) $15x + 15$
16) 7
17) 2
22) 14
23) 33
26) $3x + 2$
27) $10x - 6$
32) 20
33) -21
38) -3
39) -6

4) $-7x^2 - 2$
5) $10x^2 + 5$
6) $15x^2 + 6x$
12) $12x - 32$
13) $-12x + 4$
18) 73
19) -13
24) 0
25) 23
28) $5x + 8$
29) $-12x + 20$
34) -21
35) 4
40) -12
41) -4

7) $2x^3 + 3x^2 - 12x$
8) $2x^5 - 5x^3 - 6x^2$
9) $18x^4 + 2x^2$
14) $-3x - 6$
15) $10x + 10$
20) 6
21) 7
30) $2x - 5$
31) $24x - 5$
36) -16
37) 20
42) 2
43) -1

44)

-2 -1 0 1 2 3 4

45)

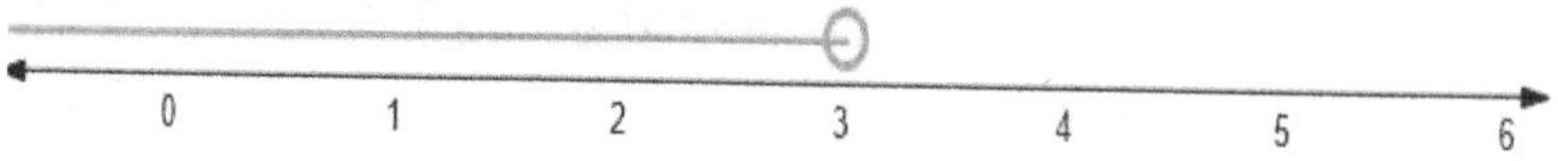

0 1 2 3 4 5 6

46)

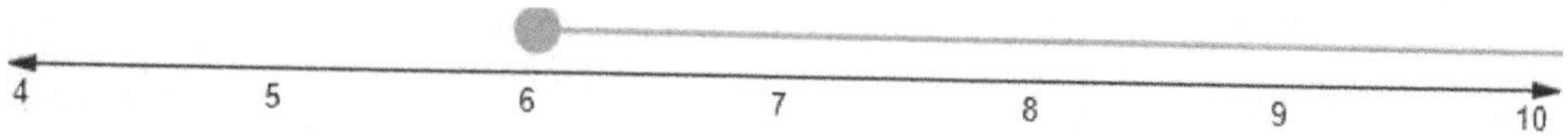

4 5 6 7 8 9 10

47)

-2 -1 0 1 2 3

48) $x \leq 7$
49) $x \leq 2$
50) $x \leq 5$

51) $x \leq 6$
52) $x \leq 2$
53) $x < 10$

Day 4: Linear Equations and Inequalities

Math Topics that you'll learn today:

- ✓ Finding Slope
- ✓ Graphing Lines Using Slope–Intercept Form
- ✓ Graphing Lines Using Standard Form
- ✓ Writing Linear Equations
- ✓ Graphing Linear Inequalities
- ✓ Finding Midpoint
- ✓ Finding Distance of Two Points

"Nature is written in mathematical language." - Galileo Galilei

Finding Slope

Step-by-step guide:

- ✓ The slope of a line represents the direction of a line on the coordinate plane.
- ✓ A coordinate plane contains two perpendicular number lines. The horizontal line is x and the vertical line is y. The point at which the two axes intersect is called the origin. An ordered pair (x, y) shows the location of a point.
- ✓ A line on coordinate plane can be drawn by connecting two points.
- ✓ To find the slope of a line, we need two points.
- ✓ The slope of a line with two points A (x_1, y_1) and B (x_2, y_2) can be found by using this formula: $\frac{y_2 - y_1}{x_2 - x_1} = \frac{rise}{\text{run}}$

Examples:

1) Find the slope of the line through these two points: $(2, -10)$ *and* $(3, 6)$.

Slope $= \frac{y_2 - y_1}{x_2 - x_1}$. Let (x_1, y_1) be $(2, -10)$ and (x_2, y_2) be $(3, 6)$. Then: slope $= \frac{y_2 - y_1}{x_2 - x_1} = \frac{6-(-10)}{3-2} = \frac{6+10}{1} = \frac{16}{1} = 16$

2) Find the slope of the line containing two points $(8, 3)$ and $(-4, 9)$.

Slope $= \frac{y_2 - y_1}{x_2 - x_1} \rightarrow (x_1, y_1) = (8,3)$ and $(x_2, y_2) = (-4, 9)$. Then: slope $= \frac{y_2 - y_1}{x_2 - x_1} = \frac{9-3}{-4-8} = \frac{6}{-12} = \frac{1}{-2} = -\frac{1}{2}$

Graphing Lines Using Slope–Intercept Form

Step-by-step guide:

- ✓ Slope-intercept form of a line: given the slope $\boldsymbol{m}$ and the y-intercept (the intersection of the line and y-axis) $\boldsymbol{b}$, then the equation of the line is:

$$\boldsymbol{y = mx + b}$$

Example: *Sketch the graph of* $y = 8x - 3$.

To graph this line, we need to find two points. When x is zero the value of y is -3. And when y is zero the value of x is $\frac{3}{8}$. $x = 0 \rightarrow y = 8(0) - 3 = -3$, $y = 0 \rightarrow 0 = 8x - 3 \rightarrow x = \frac{3}{8}$

Now, we have two points: $(0,-3)$ and $(\frac{3}{8},0)$. Find the points and graph the line. Remember that the slope of the line is 8.

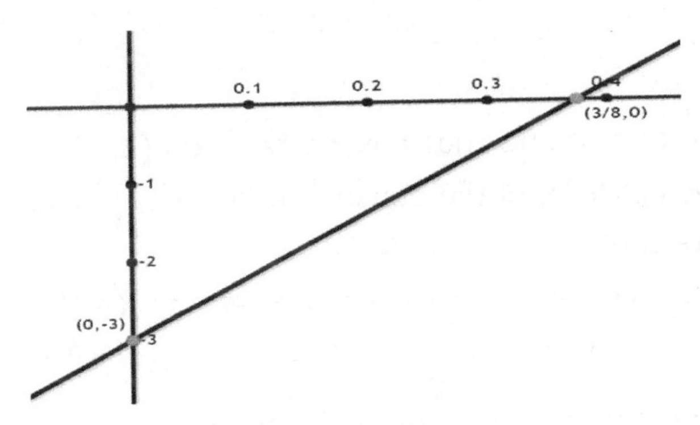

Graphing Lines Using Standard Form

Step-by-step guide:

- ✓ Find the x –intercept of the line by putting zero for $\boldsymbol{y}$.
- ✓ Find the y –intercept of the line by putting zero for the x.
- ✓ Connect these two points.

Examples:

Sketch the graph of $x - y = -5$.

First isolate y for x: $x - y = -5 \rightarrow y = x + 5$
Find the x–intercept of the line by putting zero for y.
$y = x + 5 \rightarrow x + 5 = 0 \rightarrow x = -5$

Find the y–intercept of the line by putting zero for the x.
$y = 0 + 5 \rightarrow y = 5$

Then: x–intercept: $(-5,0)$ and y–intercept: $(0,5)$

Writing Linear Equations

Step-by-step guide:

- ✓ The equation of a line: $\boldsymbol{y = mx + b}$
- ✓ Identify the slope.

- ✓ Find the y-intercept. This can be done by substituting the slope and the coordinates of a point (x, y) on the line.

Examples:

1) What is the equation of the line that passes through $(2, -2)$ and has a slope of 7?
   The general slope-intercept form of the equation of a line is $y = mx + b$, where m is the slope and b is the y-intercept.
   By substitution of the given point and given slope, we have: $-2 = (2)(7) + b$
   So, $b = -2 - 14 = -16$, and the required equation is $y = 7x - 16$.

2) Write the equation of the line through $(2, 1)$ and $(-1, 4)$.
   $Slop = \frac{y_2 - y_1}{x_2 - x_1} = \frac{4-1}{-1-2} = \frac{3}{-3} = -1 \rightarrow m = -1$
   To find the value of b, you can use either points. The answer will be the same: $y = -x + b$
   $(2, 1) \rightarrow 1 = -2 + b \rightarrow b = 3$
   $(-1, 4) \rightarrow 4 = -(-1) + b \rightarrow b = 3$
   The equation of the line is: $y = -x + 3$

Graphing Linear Inequalities

Step-by-step guide:

- ✓ First, graph the "equals" line.
- ✓ Choose a testing point. (it can be any point on both sides of the line.)
- ✓ Put the value of (x, y) of that point in the inequality. If that works, that part of the line is the solution. If the values don't work, then the other part of the line is the solution.

Examples:

Sketch the graph of $y < 2x - 3$. First, graph the line:

$y = 2x - 3$. The slope is 2 and y-intercept is -3. Then, choose a testing point. The easiest point to test is the origin: $(0, 0)$

$$(0,0) \rightarrow y < 2x - 3 \rightarrow 0 < 2(0) - 3 \rightarrow 0 < -3$$

0 is not less than -3. So, the other part of the line (on the right side) is the solution.

Finding Midpoint

Step-by-step guide:

- ✓ The middle of a line segment is its midpoint.
- ✓ The Midpoint of two endpoints A (x_1, y_1) and B (x_2, y_2) can be found using this formula: M $(\frac{x_1+x_2}{2}, \frac{y_1+y_2}{2})$

Example:

1) Find the midpoint of the line segment with the given endpoints. $(\mathbf{4}, -\mathbf{5}), (\mathbf{0}, \mathbf{9})$

Midpoint = $(\frac{x_1+x_2}{2}, \frac{y_1+y_2}{2}) \rightarrow (x_1, y_1) = (4, -5)$ and $(x_2, y_2) = (0, 9)$

Midpoint = $(\frac{4+0}{2}, \frac{-5+9}{2}) \rightarrow (\frac{4}{2}, \frac{4}{2}) \rightarrow M(2, 2)$

2) Find the midpoint of the line segment with the given endpoints. $(\mathbf{6}, \mathbf{7}), (\mathbf{4}, -\mathbf{5})$

Midpoint = $(\frac{x_1+x_2}{2}, \frac{y_1+y_2}{2}) \rightarrow (x_1, y_1) = (6, 7)$ and $(x_2, y_2) = (4, -5)$

Midpoint = $(\frac{6+4}{2}, \frac{7-5}{2}) \rightarrow (\frac{10}{2}, \frac{2}{2}) \rightarrow (5, 1)$

Finding Distance of Two Points

Step-by-step guide:

- ✓ Distance of two points A (x_1, y_1) and B (x_2, y_2): $d = \sqrt{(x_1 - x_2)^2 + (y_1 - y_2)^2}$

Examples:

1) Find the distance between of $(0, 8), (-4, 5)$.

Use distance of two points formula: $d = \sqrt{(x_1 - x_2)^2 + (y_1 - y_2)^2}$

$(x_1, y_1) = (0, 8)$ and $(x_2, y_2) = (-4, 5)$. Then: $d = \sqrt{(x_1 - x_2)^2 + (y_1 - y_2)^2} \rightarrow$

$$d = \sqrt{(0-(-4))^2 + (8-5)^2} = \sqrt{(4)^2 + (3)^2} = \sqrt{16+9} = \sqrt{25} = 5 \rightarrow d = 5$$

2) Find the distance of two points $(4, 2)$ and $(-5, -10)$.

Use distance of two points formula: $d = \sqrt{(x_1 - x_2)^2 + (y_1 - y_2)^2}$

$(x_1, y_1) = (4, 2)$, and $(x_2, y_2) = (-5, -10)$

Then: $d = \sqrt{(x_1 - x_2)^2 + (y_1 - y_2)^2} \rightarrow d = \sqrt{(4-(-5))^2 + (2-(-10))^2} =$

$\sqrt{(9)^2 + (12)^2} = \sqrt{81 + 144} = \sqrt{225} = 15$. Then: $d = 15$

Day 4 Practices

Find the slope of the line through each pair of points.

1) $(1,4),(3,8)$
2) $(-1,5),(0,6)$
3) $(5,-5),(4,-1)$
4) $(-2,-1),(0,5)$
5) $(5,1),(2,4)$
6) $(-3,5),(-2,8)$

Sketch the graph of each line. (Using Slope–Intercept Form)

7) $y = \frac{1}{2}x - 4$

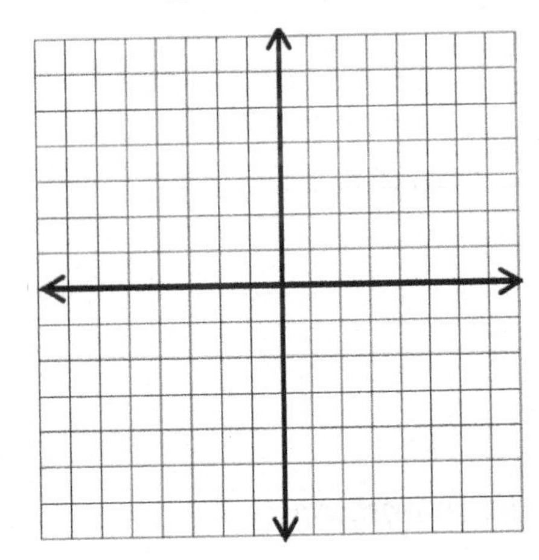

8) $y = 2x$

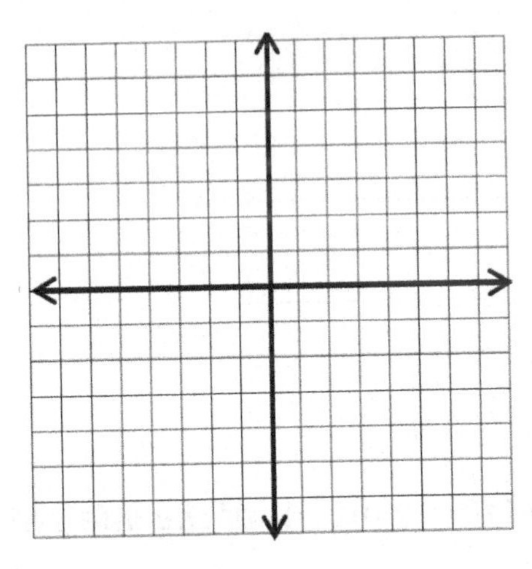

Sketch the graph of each line. (Graphing Lines Using Standard Form)

9) $y = 3x - 2$

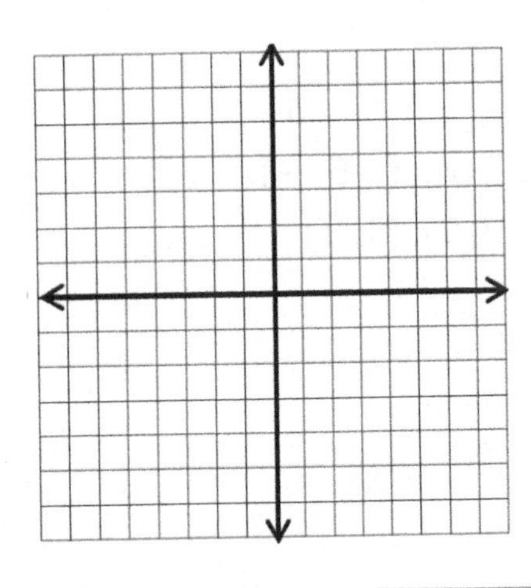

10) $y = -x + 1$

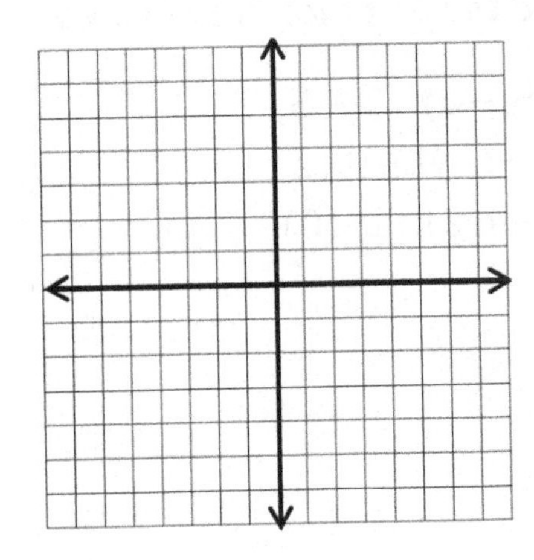

Write the equation of the line through the given points.

11) through: $(1,-2),(-2,-17)$
12) through: $(-2,1),(3,6)$
13) through: $(-2,1),(0,5)$
14) through: $(5,4),(2,1)$
15) through: $(-4,9),(3,2)$
16) through: $(1,0),(5,20)$

Sketch the graph of each linear inequality. (Graphing Linear Inequalities)

17) $2y > 6x - 2$

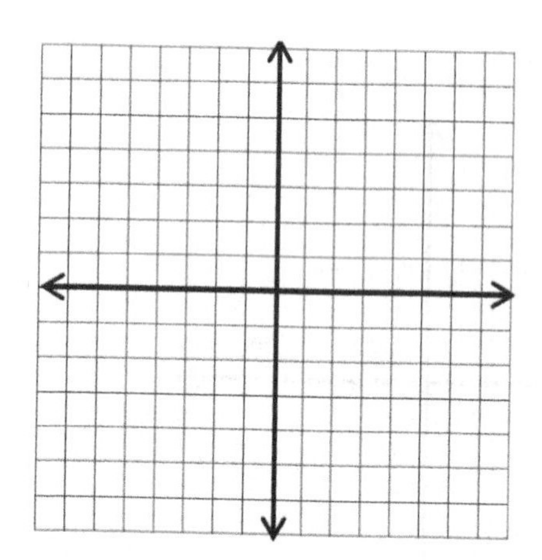

18) $3y < -3x + 12$

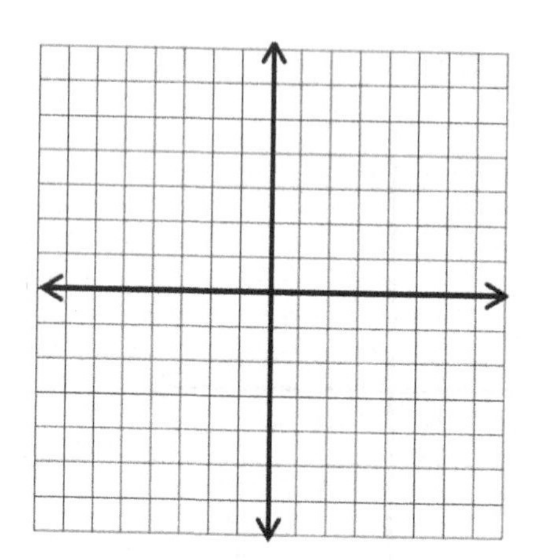

Find the midpoint of the line segment with the given endpoints.

19) $(-4,-6),(2,6)$
20) $(7,4),(-4,1)$
21) $(-4,-1),(8,3)$
22) $(-5,2),(1,6)$
23) $(3,-2),(7,-6)$
24) $(-7,-3),(5,-7)$

Find the distance between each pair of points.

25) $(5,-1),(2,-5)$
26) $(-4,-1),(0,2)$
27) $(-4,2),(2,10)$
28) $(-1,-6),(4,6)$
29) $(3,-2),(-6,-14)$
30) $(-3,0),(1,3)$

Answers

Find the slope of the line through each pair of points.

1) 2
2) 1
3) -4
4) 3
5) -1
6) 3

Sketch the graph of each line. (Using Slope–Intercept Form)

7)

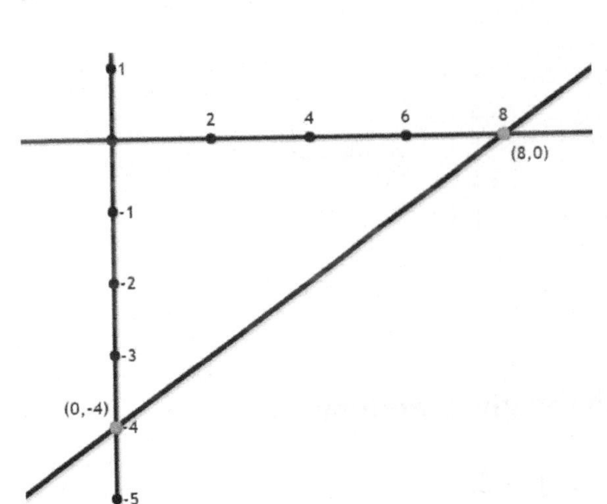

8)

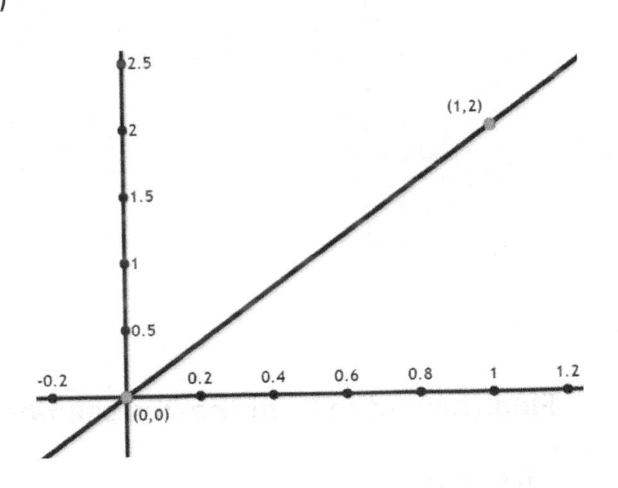

Sketch the graph of each line. (Graphing Lines Using Standard Form)

9) $y = 3x - 2$

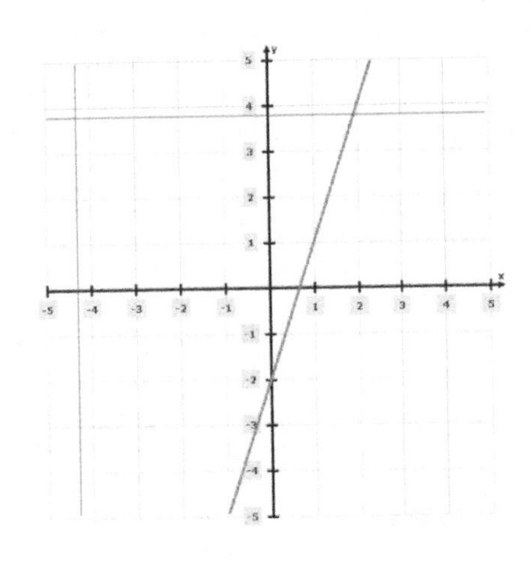

10) $y = -x + 1$

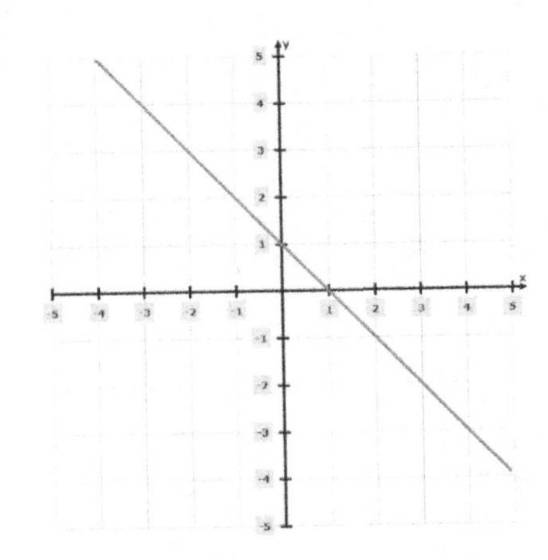

Write the equation of the line through the given points.

11) $y = 5x - 7$
12) $y = x + 3$
13) $y = 2x + 5$
14) $y = x - 1$
15) $y = -x + 5$
16) $y = 5x - 5$

Sketch the graph of each linear inequality. (Graphing Linear Inequalities)

17) $y > 3x - 1$

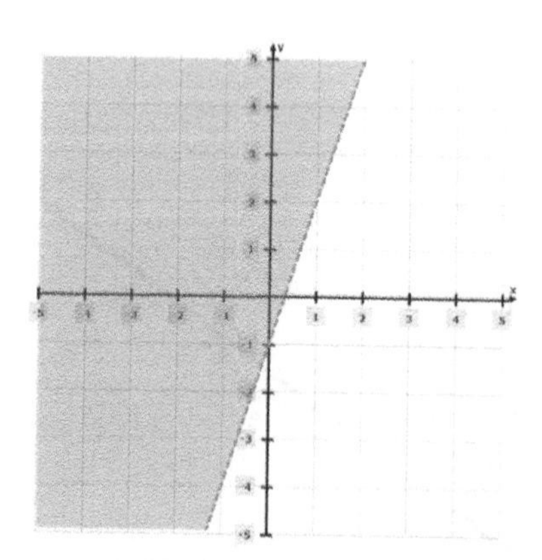

18) $y < -x + 4$

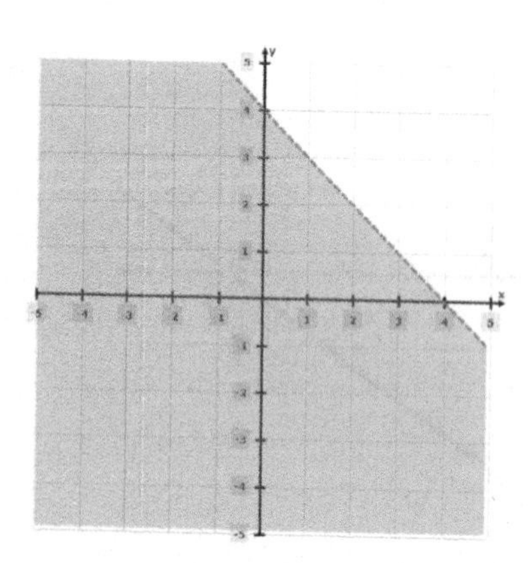

Find the midpoint of the line segment with the given endpoints.

19) $(-1, 0)$
20) $(1.5, 2.5)$
21) $(2, 1)$
22) $(-2, 4)$
23) $(5, -4)$
24) $(-1, -5)$

Find the distance between each pair of points.

25) 5
26) 5
27) 10
28) 13
29) 15
30) 5

Day 5: Monomials and Polynomials

Math Topics that you'll learn today:

- ✓ Writing Polynomials in Standard Form
- ✓ Simplifying Polynomials
- ✓ Adding and Subtracting Polynomials
- ✓ Multiplying Monomials
- ✓ Multiplying and Dividing Monomials
- ✓ Multiplying a Polynomial and a Monomial
- ✓ Multiplying Binomials
- ✓ Factoring Trinomials
- ✓ Operations with Polynomials

Mathematics is the supreme judge; from its decisions there is no appeal. - Tobias Dantzig

Simplifying Polynomials

Step-by-step guide:

- ✓ Find "like" terms. (they have same variables with same power).
- ✓ Use "FOIL". (First-Out-In-Last) for binomials:

$$(x+a)(x+b) = x^2 + (b+a)x + ab$$

- ✓ Add or Subtract "like" terms using order of operation.

Examples:

1) Simplify this expression. $4x(6x-3) =$

   Use Distributive Property: $4x(6x-3) = 24x^2 - 12x$

2) Simplify this expression. $(6x-2)(2x-3) =$

   First apply FOIL method: $(a+b)(c+d) = ac + ad + bc + bd$

   $(6x-2)(2x-3) = 12x^2 - 18x - 4x + 6$

   Now combine like terms: $12x^2 - 18x - 4x + 6 = 12x^2 - 22x + 6$

Adding and Subtracting Polynomials

Step-by-step guide:

- ✓ Adding polynomials is just a matter of combining like terms, with some order of operations considerations thrown in.
- ✓ Be careful with the minus signs, and don't confuse addition and multiplication!

Examples:

1) Simplify the expressions. $(4x^3 + 3x^4) - (x^4 - 5x^3) =$

   First use Distributive Property for $-(x^4 - 5x^3)$, $\rightarrow -(x^4 - 5x^3) = -x^4 + 5x^3$

   $(4x^3 + 3x^4) - (x^4 - 5x^3) = 4x^3 + 3x^4 - x^4 + 5x^3$

   Now combine like terms: $4x^3 + 3x^4 - x^4 + 5x^3 = 2x^4 + 9x^3$

2) Add expressions. $(2x^3 - 6) + (9x^3 - 4x^2) =$

Remove parentheses: $(2x^3 - 6) + (9x^3 - 4x^2) = 2x^3 - 6 + 9x^3 - 4x^2$

Now combine like terms: $2x^3 - 6 + 9x^3 - 4x^2 = 11x^3 - 4x^2 - 6$

Multiplying Monomials

Step-by-step guide:

✓ A monomial is a polynomial with just one term, like $2x$ or $7y$.

Examples:

1) Multiply expressions. $5a^4b^3 \times 2a^3b^2 =$

   Use this formula: $x^a \times x^b = x^{a+b}$

   $a^4 \times a^3 = a^{4+3} = a^7$ and $b^3 \times b^2 = b^{3+2} = b^5$, Then: $5a^4b^3 \times 2a^3b^2 = 10a^7b^5$

2) Multiply expressions. $-4xy^4z^2 \times 3x^2y^5z^3 =$

   Use this formula: $x^a \times x^b = x^{a+b} \rightarrow x \times x^2 = x^{1+2} = x^3$, $y^4 \times y^5 = y^{4+5} = y^9$

   and $z^2 \times z^3 = z^{2+3} = z^5$, Then: $-4xy^4z^2 \times 3x^2y^5z^3 = -12x^3y^9z^5$

Multiplying and Dividing Monomials

Step-by-step guide:

✓ When you divide two monomials you need to divide their coefficients and then divide their variables.

✓ In case of exponents with the same base, you need to subtract their powers.

✓ Exponent's rules:

$$x^a \times x^b = x^{a+b}, \quad \frac{x^a}{x^b} = x^{a-b}$$
$$\frac{1}{x^b} = x^{-b}, \quad (x^a)^b = x^{a \times b}$$
$$(xy)^a = x^a \times y^a$$

Examples:

1) Multiply expressions. $(-3x^7)(4x^3) =$
   Use this formula: $x^a \times x^b = x^{a+b} \rightarrow x^7 \times x^3 = x^{10}$
   Then: $(-3x^7)(4x^3) = -12x^{10}$

2) Dividing expressions. $\frac{18x^2y^5}{2xy^4} =$

Use this formula: $\frac{x^a}{x^b} = x^{a-b}$, $\frac{x^2}{x} = x^{2-1} = x$ and $\frac{y^5}{y^4} = y^{5-4} = y$

Then: $\frac{18x^2y^5}{2xy^4} = 9xy$

Multiplying a Polynomial and a Monomial

Step-by-step guide:

- ✓ When multiplying monomials, use the product rule for exponents.
- ✓ When multiplying a monomial by a polynomial, use the distributive property.

$$a \times (b + c) = a \times b + a \times c$$

Examples:

1) Multiply expressions. $-4x(5x + 9) =$

Use Distributive Property: $-4x(5x + 9) = -20x^2 - 36x$

2) Multiply expressions. $2x(6x^2 - 3y^2) =$

Use Distributive Property: $2x(6x^2 - 3y^2) = 12x^3 - 6xy^2$

Multiplying Binomials

Step-by-step guide:

- ✓ Use "FOIL". (First-Out-In-Last)

$$(x + a)(x + b) = x^2 + (b + a)x + ab$$

Examples:

1) Multiply Binomials. $(x - 2)(x + 2) =$

Use "FOIL". (First–Out–In–Last): $(x - 2)(x + 2) = x^2 + 2x - 2x - 4$

Then simplify: $x^2 + 2x - 2x - 4 = x^2 - 4$

2) Multiply Binomials. $(x+5)(x-2) =$

Use "FOIL". (First–Out–In–Last):

$(x+5)(x-2) = x^2 - 2x + 5x - 10$

Then simplify: $x^2 - 2x + 5x - 10 = x^2 + 3x - 10$

Factoring Trinomials

Step-by-step guide:

✓ "FOIL": $(x+a)(x+b) = x^2 + (b+a)x + ab$

✓ "Difference of Squares": $a^2 - b^2 = (a+b)(a-b)$

$$a^2 + 2ab + b^2 = (a+b)(a+b)$$

$$a^2 - 2ab + b^2 = (a-b)(a-b)$$

✓ "Reverse FOIL": $x^2 + (b+a)x + ab = (x+a)(x+b)$

Examples:

1) Factor this trinomial. $x^2 - 2x - 8 =$

Break the expression into groups: $(x^2 + 2x) + (-4x - 8)$

Now factor out x from $x^2 + 2x$: $x(x+2)$ and factor out -4 from $-4x - 8$: $-4(x+2)$

Then: $= x(x+2) - 4(x+2)$, now factor out like term: $x + 2$

Then: $(x+2)(x-4)$

2) Factor this trinomial. $x^2 - 6x + 8 =$

Break the expression into groups: $(x^2 - 2x) + (-4x + 8)$

Now factor out x from $x^2 - 2x$: $x(x-2)$, and factor out -4 from $-4x + 8$: $-4(x-2)$

Then: $= x(x-2) - 4(x-2)$, now factor out like term: $x - 2$

Then: $(x-2)(x-4)$

Operations with Polynomials

Step-by-step guide:

- ✓ When multiplying a monomial by a polynomial, use the distributive property.

$$a \times (b + c) = a \times b + a \times c = ab + ac$$

Examples:

1) Multiply. $5(2x - 6) =$

   Use the distributive property: $5(2x - 6) = 10x - 30$

2) Multiply. $2x(6x + 2) =$

   Use the distributive property: $2x(6 + 2) = 12x^2 + 4x$

Day 5 Practices

Simplify each polynomial.

1) $5(2x-10)=$

2) $2x(4x-2)=$

3) $4x(5x-3)=$

4) $3x(7x+3)=$

5) $4x(8x-4)=$

6) $5x(5x+4)=$

Add or subtract polynomials.

7) $(-x^2-2)+(2x^2+1)=$

8) $(2x^2+3)-(3-4x^2)=$

9) $(2x^3+3x^2)-(x^3+8)=$

10) $(4x^3-x^2)+(3x^2-5x)=$

11) $(7x^3+9x)-(3x^3+2)=$

12) $(2x^3-2)+(2x^3+2)=$

Simplify each expression. (Multiplying Monomials)

13) $4u^7\times(-2u^5)=$

14) $(-2p^7)\times(-3p^2)=$

15) $3xy^2z^3\times 2z^2=$

16) $5u^5t\times 3ut^2=$

17) $(-9a^6)\times(-5a^2b^4)=$

18) $-2a^3b^2\times 4a^2b=$

Simplify each expression. (Multiplying and Dividing Monomials)

19) $(3x^7y^2)(16x^5y^4)=$

20) $(4x^4y^6)(7x^3y^4)=$

21) $(7x^2y^9)(12x^9y^{12})=$

22) $\frac{12x^6y^8}{4x^4y^2}=$

23) $\frac{52x^9y^5}{4x^3y^4}=$

24) $\frac{80x^{12}y^9}{10x^6y^7}=$

Find each product. (Multiplying a Polynomial and a Monomial)

25) $3x(9x + 2y) =$

26) $6x(x + 2y) =$

27) $9x(2x + 4y) =$

28) $12x(3x + 9y) =$

29) $11x(2x - 11y) =$

30) $2x(6x - 6y) =$

Find each product. (Multiplying Binomials)

31) $(x + 2)(x + 2) =$

32) $(x - 3)(x + 2) =$

33) $(x - 2)(x - 4) =$

34) $(x + 3)(x + 2) =$

35) $(x - 4)(x - 5) =$

36) $(x + 5)(x + 2) =$

Factor each trinomial.

37) $x^2 + 8x + 15 =$

38) $x^2 - 5x + 6 =$

39) $x^2 + 6x + 8 =$

40) $x^2 - 8x + 16 =$

41) $x^2 - 7x + 12 =$

42) $x^2 + 11x + 18 =$

Find each product. (Operations with Polynomials)

43) $9(6x + 2) =$

44) $8(3x + 7) =$

45) $5(6x - 1) =$

46) $-3(8x - 3) =$

47) $3x^2(6x - 5) =$

48) $5x^2(7x - 2) =$

Answers

1) $10x - 50$
2) $8x^2 - 4x$
3) $20x^2 - 12x$
4) $21x^2 + 9x$
5) $32x^2 - 16x$
6) $25x^2 + 20x$
7) $x^2 - 1$
8) $6x^2$
9) $x^3 + 3x^2 - 8$
10) $4x^3 + 2x^2 - 5x$
11) $4x^3 + 9x - 2$
12) $4x^3$
13) $-8u^{12}$
14) $6p^9$
15) $6xy^2z^5$
16) $15u^6t^3$
17) $45a^8b^4$
18) $-8a^5b^3$
19) $48x^{12}y^6$
20) $28x^7y^{10}$
21) $84x^{11}y^{21}$
22) $3x^2y^6$
23) $13x^6y$
24) $8x^6y^2$
25) $27x^2 + 6xy$
26) $6x^2 + 12xy$
27) $18x^2 + 36xy$
28) $36x^2 + 108xy$
29) $22x^2 - 121xy$
30) $12x^2 - 12xy$
31) $x^2 + 4x + 4$
32) $x^2 - x - 6$
33) $x^2 - 6x + 8$
34) $x^2 + 5x + 6$
35) $x^2 - 9x + 20$
36) $x^2 + 7x + 10$
37) $(x + 3)(x + 5)$
38) $(x - 2)(x - 3)$
39) $(x + 4)(x + 2)$
40) $(x - 4)(x - 4)$
41) $(x - 3)(x - 4)$
42) $(x + 2)(x + 9)$
43) $54x + 18$
44) $24x + 56$
45) $30x - 5$
46) $-24x + 9$
47) $18x^3 - 15x^2$
48) $35x^3 - 10x^2$

Day 6: Geometry and Statistics

Math Topics that you'll learn today:

- ✓ The Pythagorean Theorem
- ✓ Triangles
- ✓ Polygons
- ✓ Circles
- ✓ Trapezoids
- ✓ Cubes
- ✓ Rectangle Prisms
- ✓ Cylinder
- ✓ Mean, Median, Mode, and Range of the Given Data
- ✓ Bar Graph
- ✓ Box and Whisker Plots
- ✓ Stem– And– Leaf Plot
- ✓ Pie Graph
- ✓ Probability

Mathematics is like checkers in being suitable for the young, not too difficult, amusing, and without peril to the state. ~ Plato

The Pythagorean Theorem

Step-by-step guide:

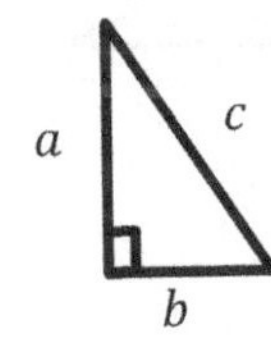

- ✓ In any right triangle: $a^2 + b^2 = c^2$

Examples:

1) Find the missing length.

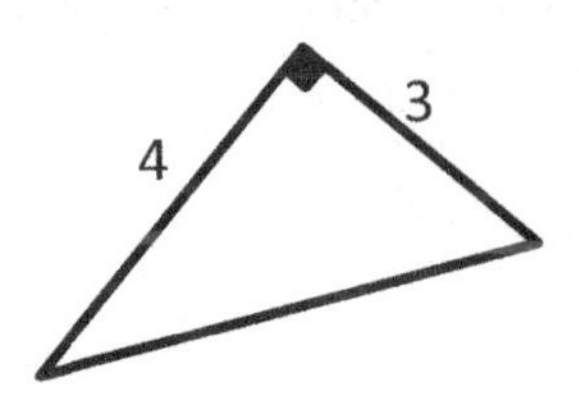

Use Pythagorean Theorem: $a^2 + b^2 = c^2$

Then: $a^2 + b^2 = c^2 \rightarrow 3^2 + 4^2 = c^2 \rightarrow 9 + 16 = c^2$

$c^2 = 25 \rightarrow c = 5$

2) Right triangle ABC has two legs of lengths 6 cm (AB) and 8 cm (AC). What is the length of the third side (BC)?

Use Pythagorean Theorem: $a^2 + b^2 = c^2$

Then: $a^2 + b^2 = c^2 \rightarrow 6^2 + 8^2 = c^2 \rightarrow 36 + 64 = c^2$

$c^2 = 100 \rightarrow c = 10$

Triangles

Step-by-step guide:

- ✓ In any triangle the sum of all angles is 180 degrees.
- ✓ Area of a triangle = $\frac{1}{2}$ $(base \times height)$
- ✓ All angles in a triangle sum up to 180 degrees.

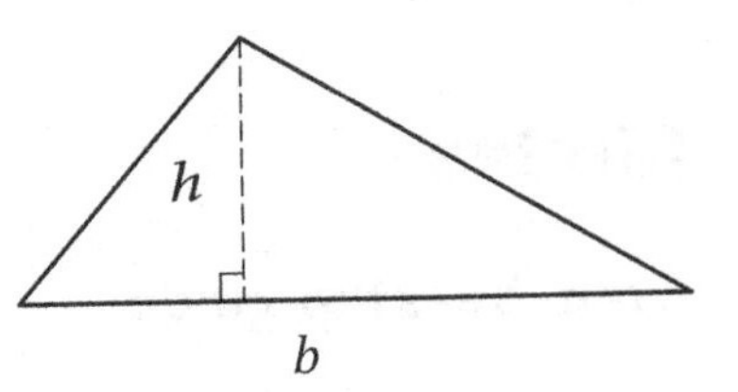

Examples:

What is the area of triangles?

1)

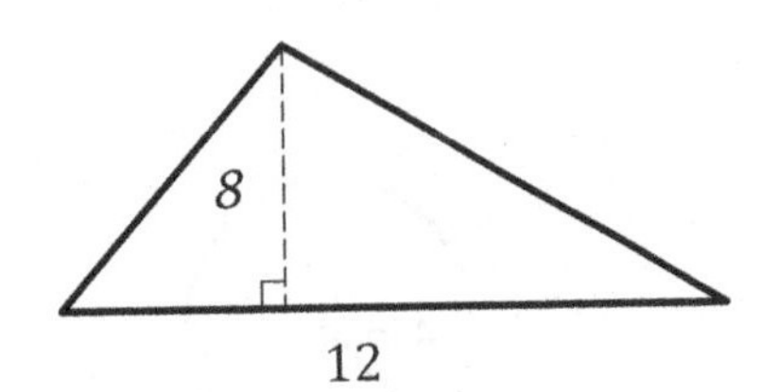

Solution:

Use the are formula: Area $= \frac{1}{2}$ $(base \times height)$

$base = 12$ and $height = 8$

Area $= \frac{1}{2}(12 \times 8) = \frac{1}{2}(96) = 48$

2)

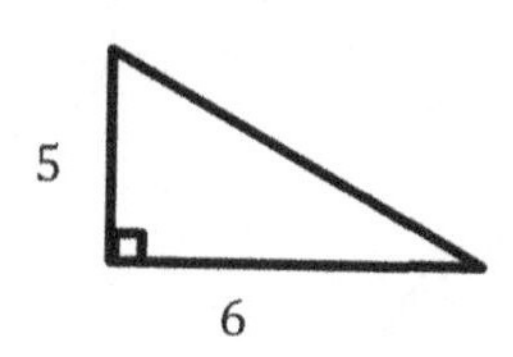

Solution:

Use the are formula: Area $= \frac{1}{2}$ $(base \times height)$

$base = 6$ and $height = 5$

Area $= \frac{1}{2}(5 \times 6) = \frac{30}{2} = 15$

Polygons

Step-by-step guide:

Perimeter of a square $= 4 \times side = 4s$

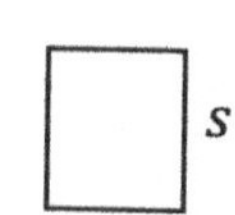

Perimeter of a rectangle

$= 2(width + length)$

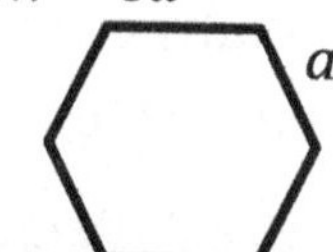

Perimeter of trapezoid

$= a + b + c + d$

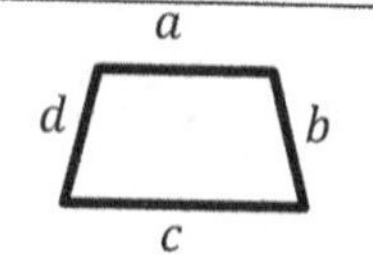

Perimeter of a regular hexagon $= 6a$

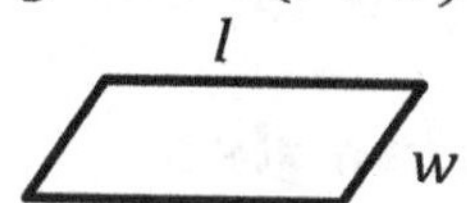

Example: Find the perimeter of following regular hexagon.

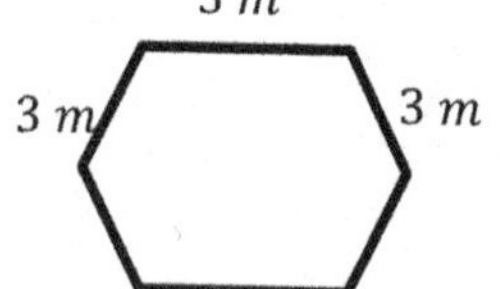

Perimeter of Pentagon $= 6a$

Perimeter of Pentagon $= 6a = 6 \times 3 = 18m$

Perimeter of a parallelogram $= 2(l + w)$

l

w

Circles

Step-by-step guide:

- ✓ In a circle, variable r is usually used for the radius and d for diameter and π is about 3.14.
- ✓ $Area\ of\ a\ circle = \pi r^2$
- ✓ $Circumference\ of\ a\ circle = 2\pi r$

Examples:

1) Find the area of the circle.

Use area formula: $Area = \pi r^2$,

$r = 4$ then: $Area = \pi(4)^2 = 16\pi$, $\pi = 3.14$ then:

$Area = 16 \times 3.14 = 50.24$

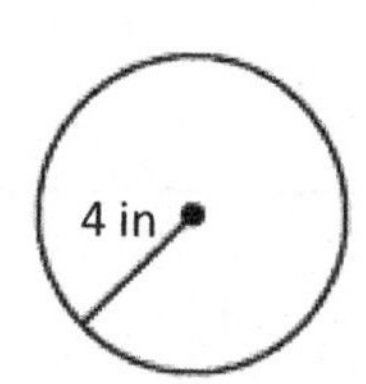

2) Find the Circumference of the circle.

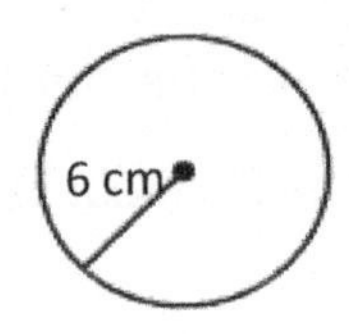

Use Circumference formula: $Circumference = 2\pi r$
$r = 6$, then: $Circumference = 2\pi(6) = 12\pi$
$\pi = 3.14$ then: $Circumference = 12 \times 3.14 = 37.68$
$(\pi = 3.14)$

Trapezoids

Step-by-step guide:

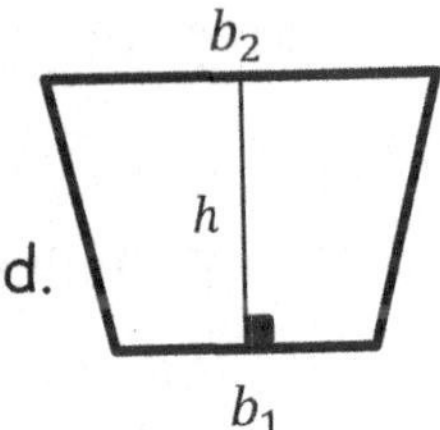

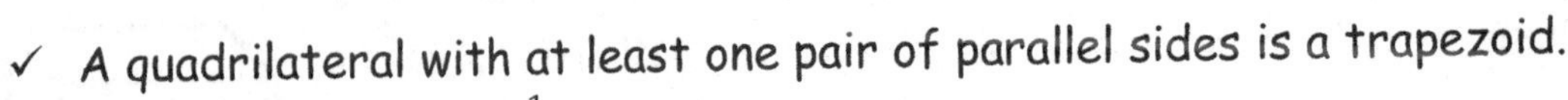

- ✓ A quadrilateral with at least one pair of parallel sides is a trapezoid.
- ✓ Area of a trapezoid $= \frac{1}{2}h(b_1 + b_2)$

Example:

Calculate the area of the trapezoid.

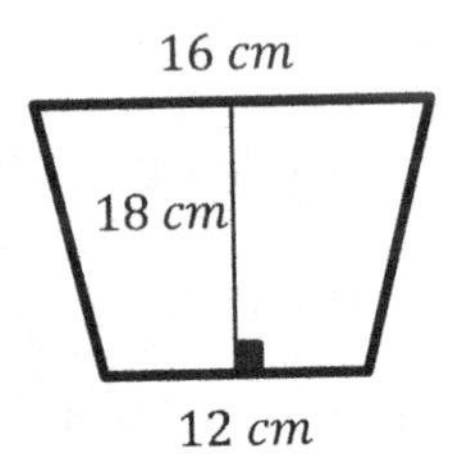

Use area formula: $A = \frac{1}{2}h(b_1 + b_2)$

$b_1 = 12$, $b_2 = 16$ and $h = 18$

Then: $A = \frac{1}{2}18(12 + 16) = 9(28) = 252\ cm^2$

Cubes

Step-by-step guide:

- ✓ A cube is a three-dimensional solid object bounded by six square sides.
- ✓ Volume is the measure of the amount of space inside of a solid figure, like a cube, ball, cylinder or pyramid.
- ✓ Volume of a cube $= (\boldsymbol{one\ side})^3$
- ✓ surface area of cube $= \boldsymbol{6} \times (\boldsymbol{one\ side})^2$

Example:

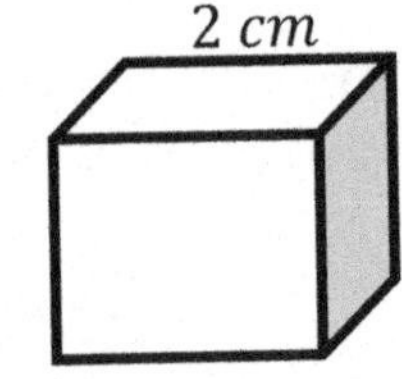

Find the volume and surface area of this cube.

Use volume formula: $volume = (one\ side)^3$

Then: $volume = (one\ side)^3 = (2)^3 = 8\ cm^3$

Use surface area formula: $surface\ area\ of\ cube$: $6(one\ side)^2 = 6(2)^2 = 6(4) = 24\ cm^2$

Rectangular Prisms

Step-by-step guide:

- ✓ A solid 3-dimensional object which has six rectangular faces.
- ✓ Volume of a Rectangular prism = $\boldsymbol{Length} \times \boldsymbol{Width} \times \boldsymbol{Height}$

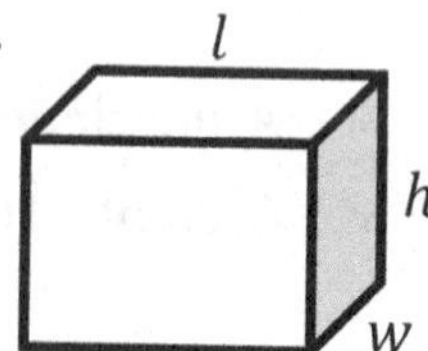

$Volume = l \times w \times h$ $\quad$ $Surface\ area = 2(wh + lw + lh)$

Example:

Find the volume and surface area of rectangular prism.

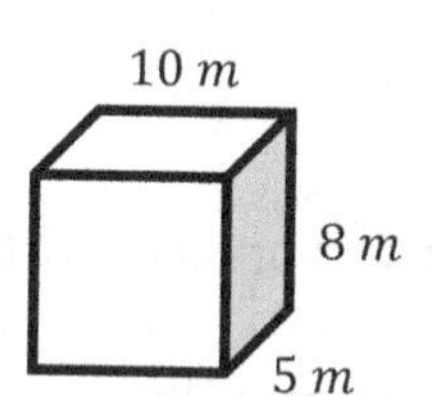

Use volume formula: $Volume = l \times w \times h$

Then: $Volume = 10 \times 5 \times 8 = 400\ m^3$

Use surface area formula: $Surface\ area = 2(wh + lw + lh)$

Then: $Surface\ area = 2(5 \times 8 + 10 \times 5 + 10 \times 8) = 2(40 + 50 + 80) = 340\ m^2$

Cylinder

Step-by-step guide:

- ✓ A cylinder is a solid geometric figure with straight parallel sides and a circular or oval cross section.
- ✓ $Volume\ of\ Cylinder\ Formula = \pi\ (radius)^2 \times height$,($\pi = 3.14$)
- ✓ $Surface\ area\ of\ a\ cylinder = 2\pi r^2 + 2\pi rh$

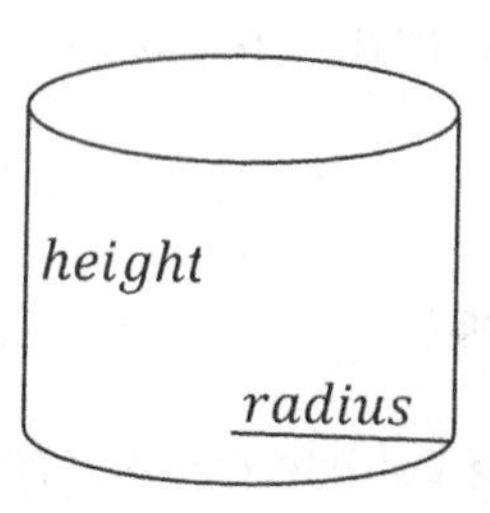

Example:

Find the volume and Surface area of the follow Cylinder.

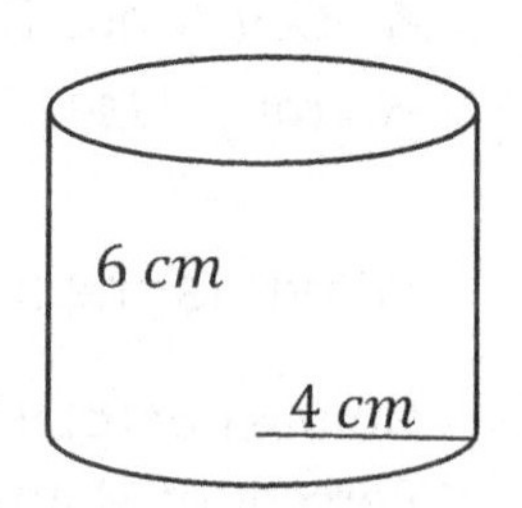

Use volume formula: $Volume = \pi(radius)^2 \times height$
Then: $Volume = \pi(4)^2 \times 6 = \pi 16 \times 6 = 96\pi$
$\pi = 3.14$ then: $Volume = 96\pi = 301.44$
Use surface area formula: $Surface\ area = 2\pi r^2 + 2\pi rh$
Then: $= 2\pi(4)^2 + 2\pi(4)(6) = 2\pi(16) + 2\pi(24) = 32\pi + 48\pi = 80\pi$
$\pi = 3.14$ then: $Surface\ area = 80 \times 3.14 = 251.2$

Mean, Median, Mode, and Range of the Given Data

Step-by-step guide:

- ✓ Mean: $\frac{\text{sum of the data}}{\text{total number of data entires}}$
- ✓ Mode: value in the list that appears most often
- ✓ Range: the difference of largest value and smallest value in the list

Examples:

1) What is the median of these numbers? $4, 9, 13, 8, 15, 18, 5$

   Write the numbers in order: $4, 5, 8, 9, 13, 15, 18$

   Median is the number in the middle. Therefore, the median is 9.

2) What is the mode of these numbers? $22, 16, 12, 9, 7, 6, 4, 6$

   Mode: value in the list that appears most often
   Therefore: mode is 6

Pie Graph

Step-by-step guide:

- ✓ A Pie Chart is a circle chart divided into sectors; each sector represents the relative size of each value.

Example:

A library has 840 books that include Mathematics, Physics, Chemistry, English and History. Use following graph to answer question.

What is the number of Mathematics books?

Number of total books $= 840$,
Percent of Mathematics books $= 30\% = 0.30$
Then: $0.30 \times 840 = 252$

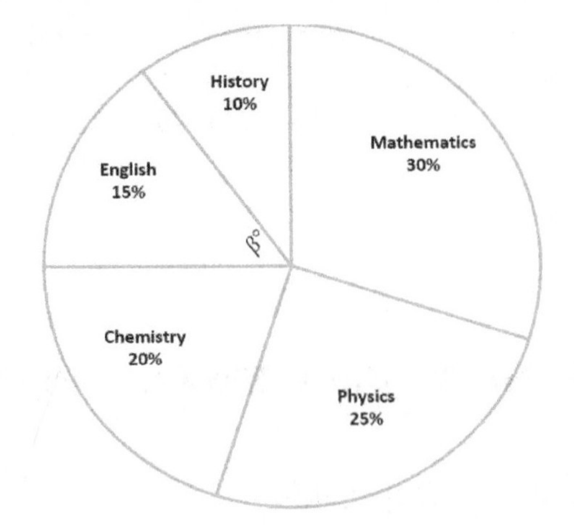

Probability Problems

Step-by-step guide:

- ✓ Probability is the likelihood of something happening in the future. It is expressed as a number between zero (can never happen) to 1 (will always happen).
- ✓ Probability can be expressed as a fraction, a decimal, or a percent.

Examples:

1) If there are 8 red balls and 12 blue balls in a basket, what is the probability that John will pick out a red ball from the basket?
   There are 8 red ball and 20 are total number of balls. Therefore, probability that John will pick out a red ball from the basket is 8 out of 20 or $\frac{8}{8+12} = \frac{8}{20} = \frac{2}{5}$.
2) A bag contains 18 balls: two green, five black, eight blue, a brown, a red and one white. If 17 balls are removed from the bag at random, what is the probability that a brown ball has been removed?
   If 17 balls are removed from the bag at random, there will be one ball in the bag.
   The probability of choosing a brown ball is 1 out of 18. Therefore, the probability of not choosing a brown ball is 17 out of 18 and the probability of having not a brown ball after removing 17 balls is the same.

Day 6 Practices

Find the missing side?

1)

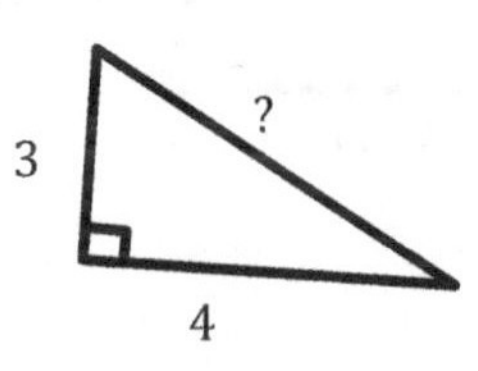

2)

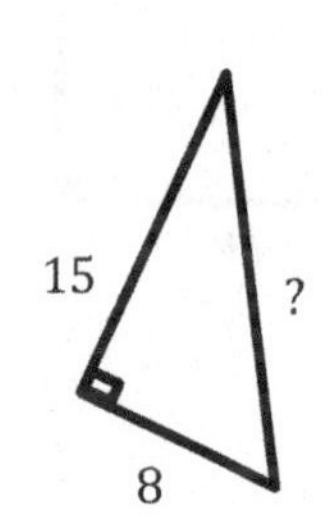

3)

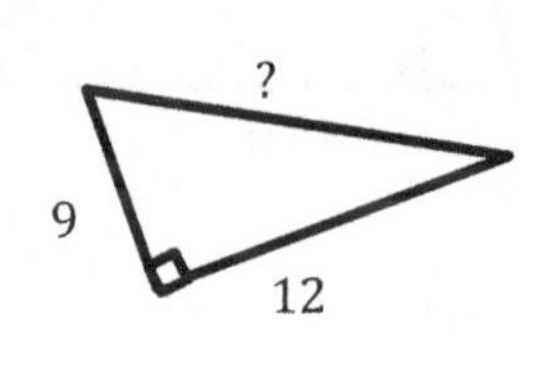

4)

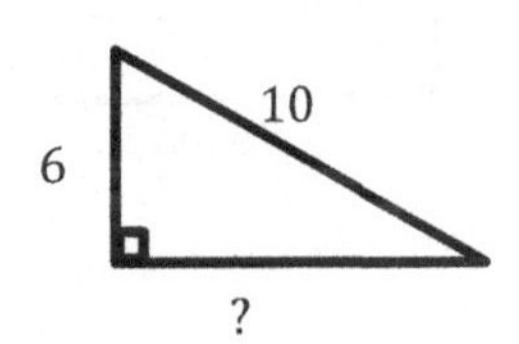

Find the measure of the unknown angle in each triangle.

5)

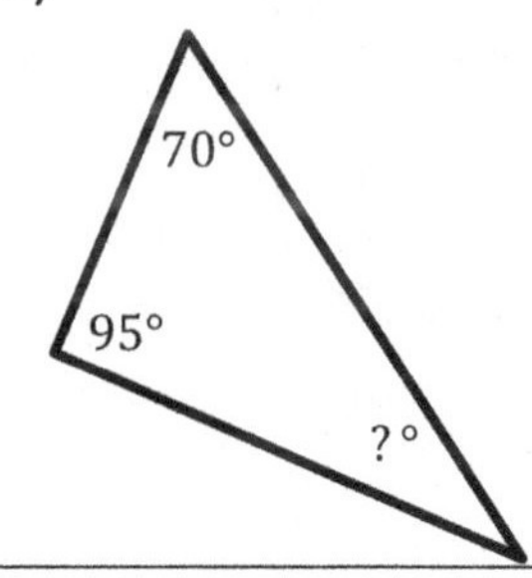

6)

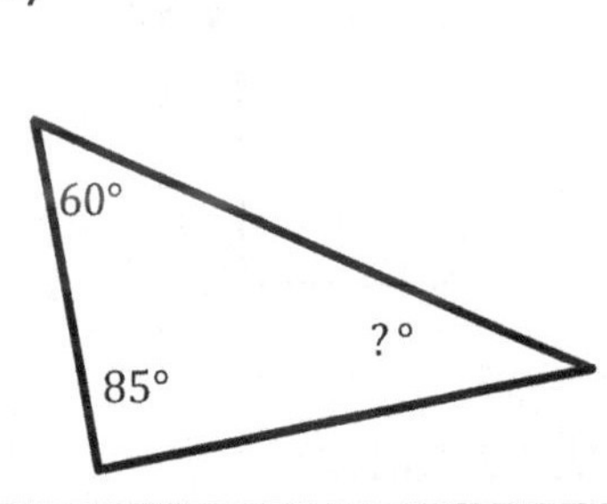

7)

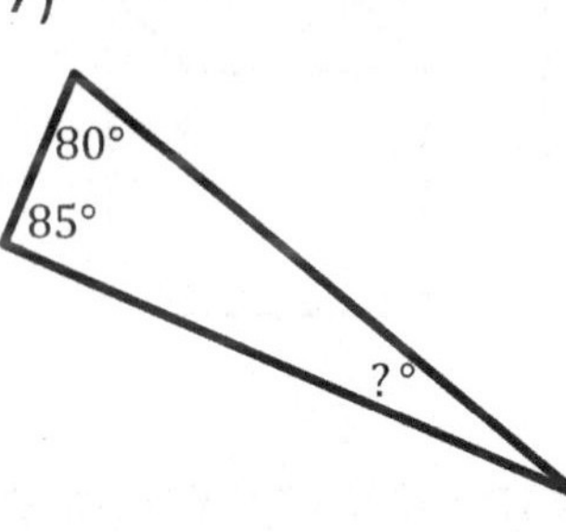

8)

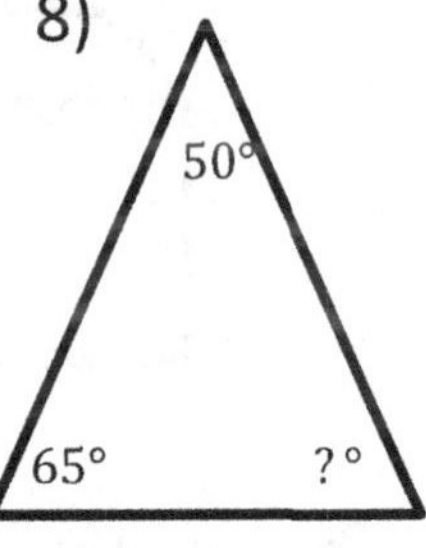

Find the perimeter of each shape.

9)

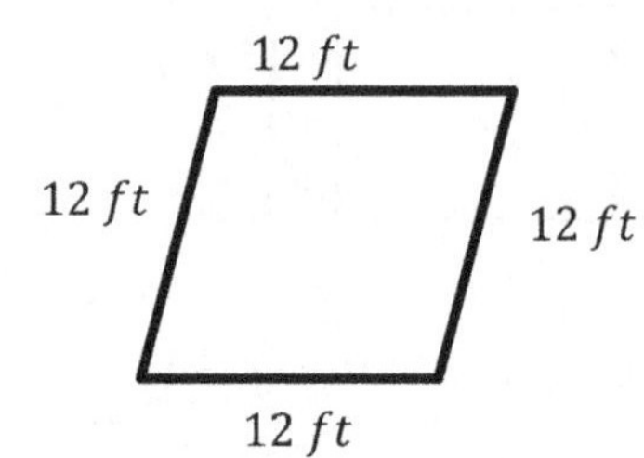

10)

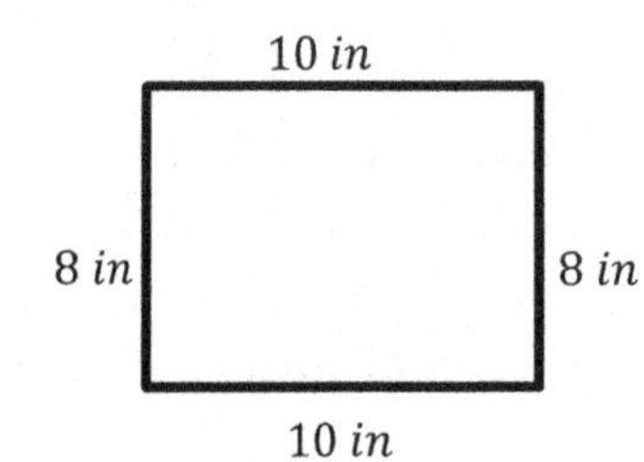

11)

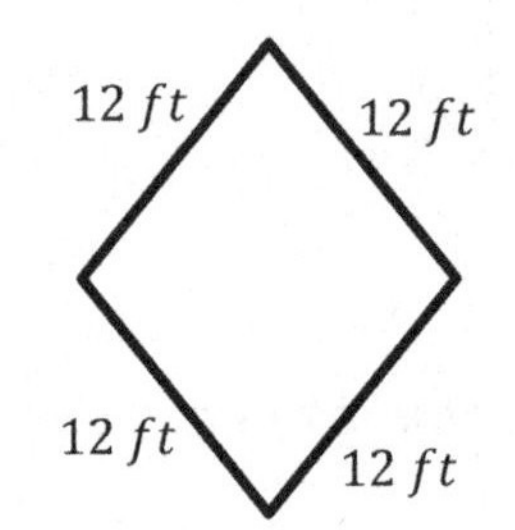

12)

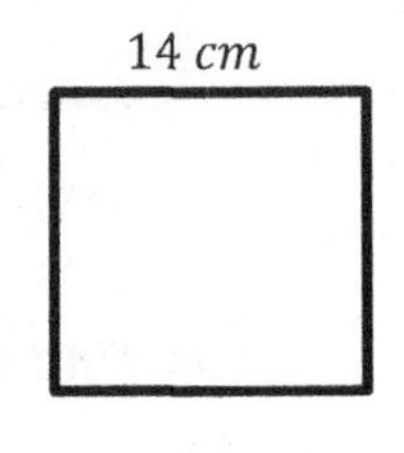

Complete the table below. ($\pi = 3.14$)

13)

	Radius	Diameter	Circumference	Area
Circle 1	4 *inches*	8 *inches*	25.12 *inches*	50.24 *square inches*
Circle 2		12 *meters*		
Circle 3				12.56 *square ft*
Circle 4			18.84 miles	

Find the area of each trapezoid.

14)

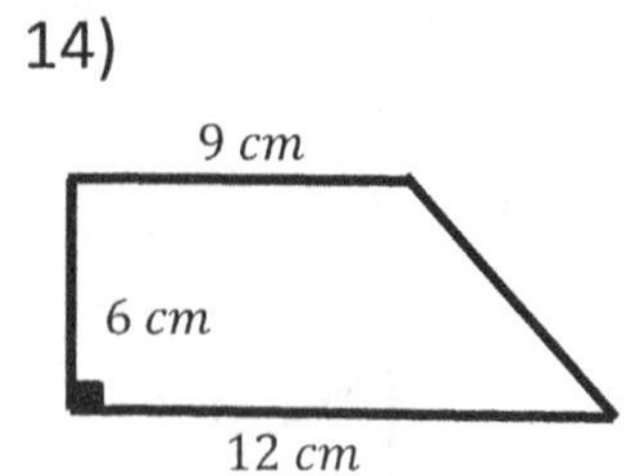

15)

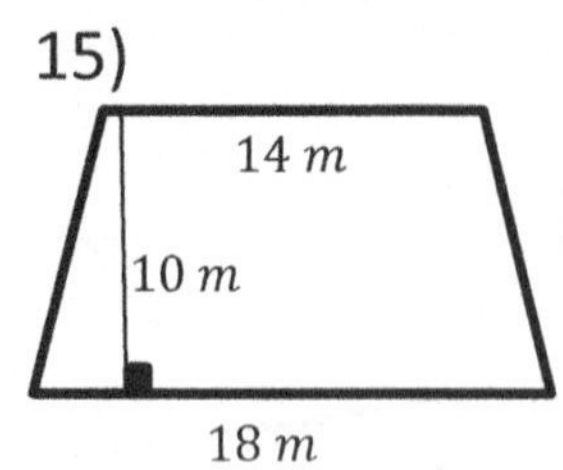

16)

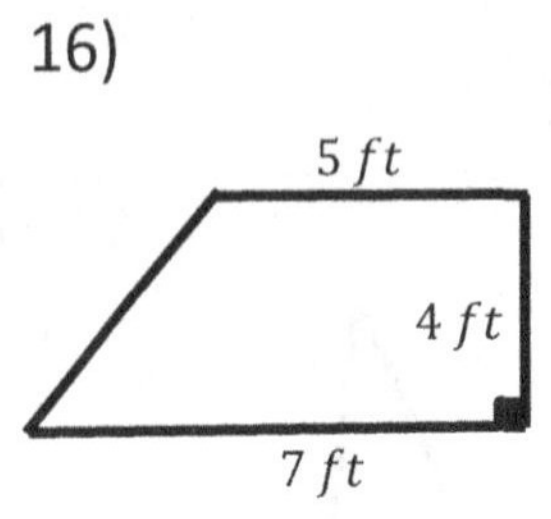

17)

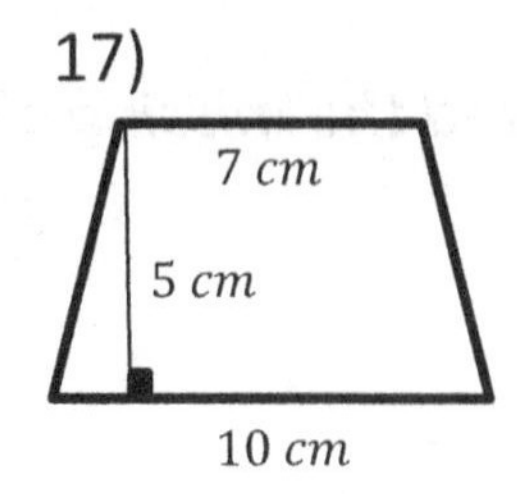

Find the volume of each cube.

18)

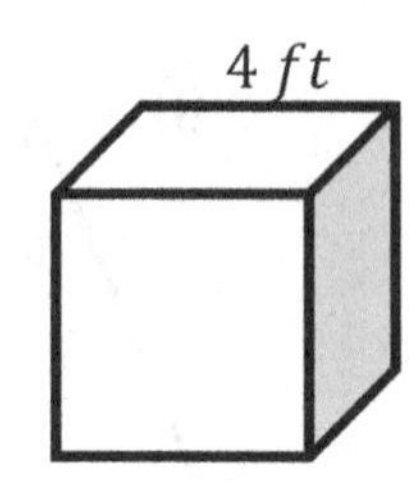

19)

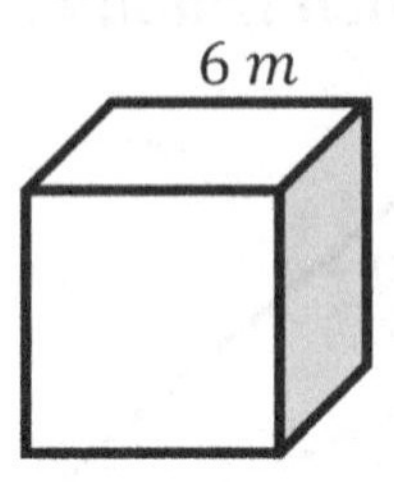

20)

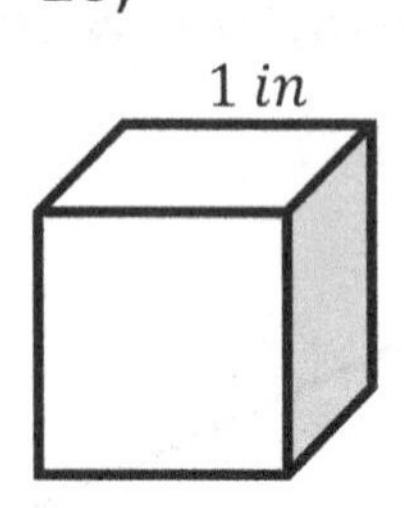

21)

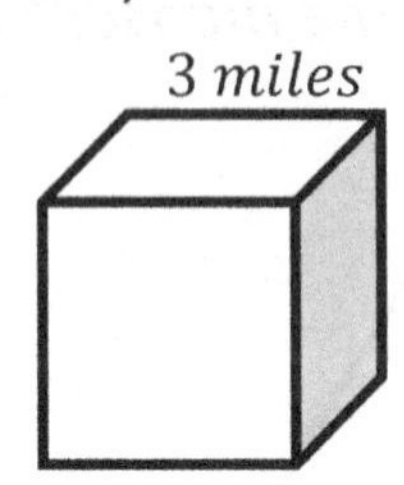

Find the volume of each Rectangular Prism.

22)

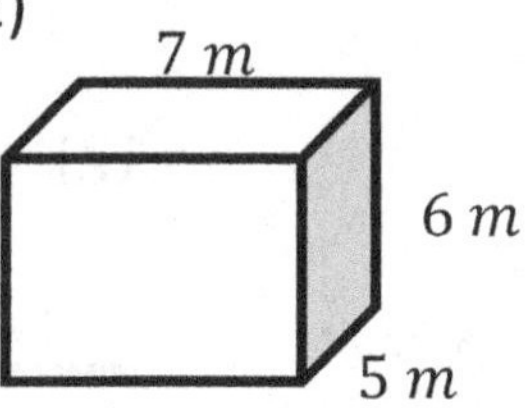

23)

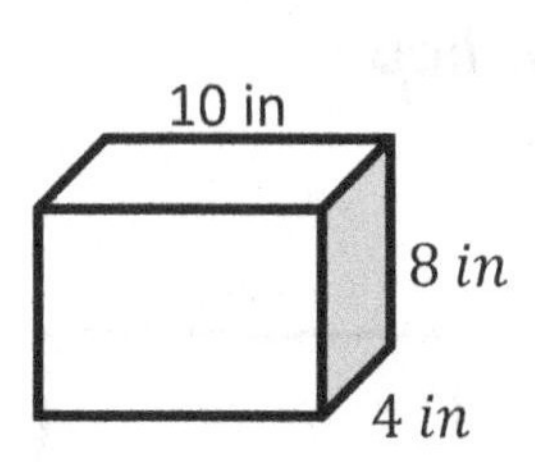

24)

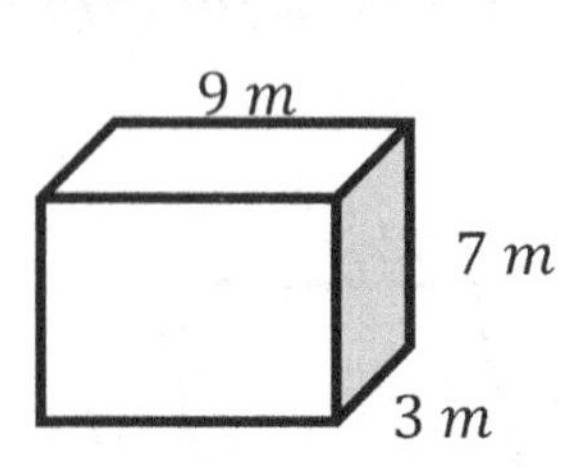

Find the volume of each Cylinder. Round your answer to the nearest tenth. ($\pi = 3.14$)

25)

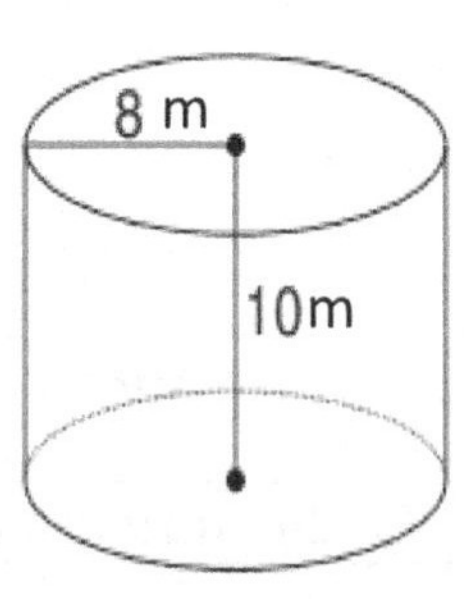

26)

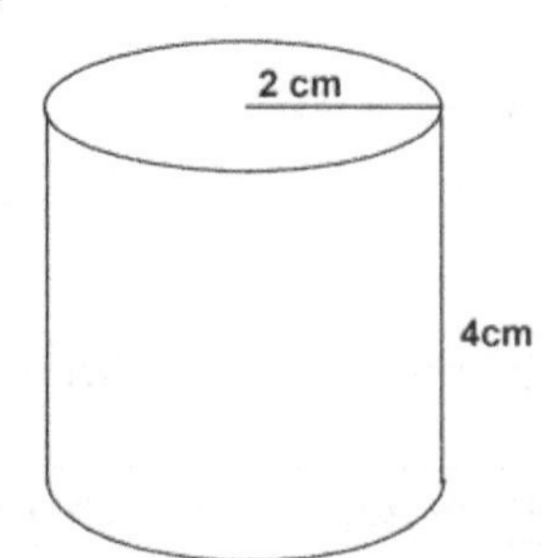

27)

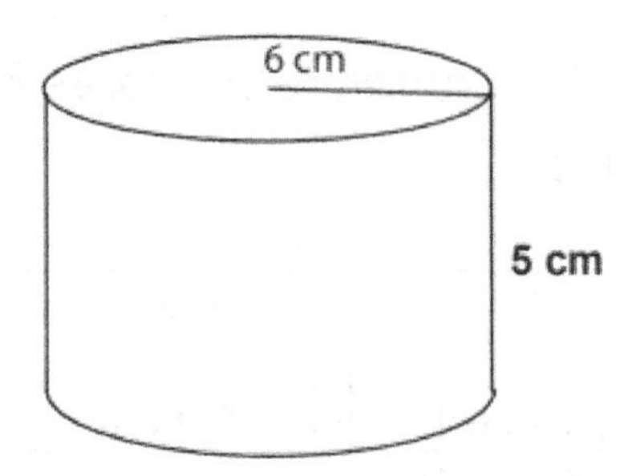

Solve.

28) In a javelin throw competition, five athletics score 56, 58, 63, 57 and 61 meters. What are their Mean and Median? ________________

The circle graph below shows all Jason's expenses for last month. Jason spent $300 on his bills last month.

29) How much did Jason spend on his car last month? ________

30) How much did Jason spend for foods last month? ________

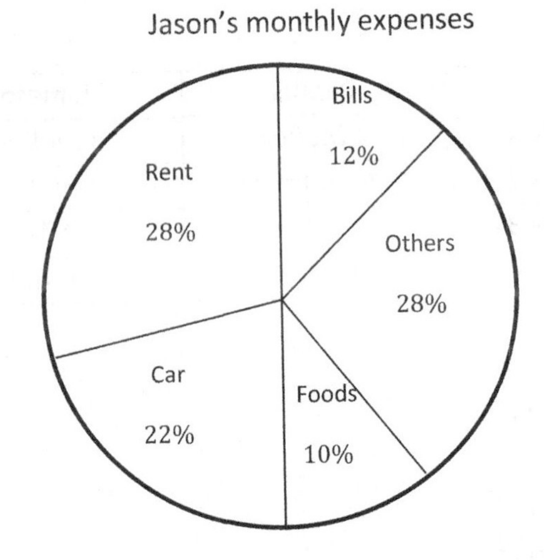

Solve.

31) Bag A contains 9 red marbles and 3 green marbles. Bag B contains 9 black marbles and 6 orange marbles. What is the probability of selecting a green marble at random from bag A? What is the probability of selecting a black marble at random from Bag B? ________ ____________

Answers

1) 5
2) 17
3) 15
4) 8

5) 15°
6) 35°
7) 15°
8) 65°

9) 48 ft
10) 36 in
11) 48 ft
12) 56 cm

13)

	Radius	Diameter	Circumference	Area
Circle 1	4 *inches*	8 *inches*	25.12 *inches*	50.24 *square inches*
Circle 2	6 *meters*	12 *meters*	37.68 *meters*	113.04 *square meters*
Circle 3	2 ft	4 ft	12.56 ft	12.56 *square* ft
Circle 4	3 *miles*	6 *miles*	18.84 *miles*	28.26 *square miles*

14) 63 cm^2
15) 160 m^2
16) 24 ft^2
17) 42.5 cm^2

18) 64 ft^3
19) 216 m^3
20) 1 in^3
21) 27 $miles^3$

22) 210 m^3
23) 320 in^3
24) 189 m^3

25) 2,009.6 m^3
26) 50.2 cm^3
27) 565.2 cm^3
28) *Mean*: 59, *Median*: 58
29) \$550
30) \$250
31) $\frac{1}{4}, \frac{3}{5}$

Day 7: Geometry and Statistics

Math Topics that you'll learn today:

- ✓ Systems of Equations
- ✓ Quadratic Equation
- ✓ Graphing Quadratic Functions
- ✓ Quadratic Inequalities
- ✓ Graphing Quadratic inequalities
- ✓ Adding and subtracting complex numbers
- ✓ Multiplying and dividing complex numbers
- ✓ Rationalizing Imaginary Denominators
- ✓ Function Notation
- ✓ Adding and Subtracting Functions
- ✓ Multiplying and Dividing Functions
- ✓ Composition of Functions
- ✓ Trig Ratios of General Angles
- ✓ Angles and Angle Measure
- ✓ Evaluating Trigonometric Function

Systems of Equations

Step-by-step guide:

- ✓ A system of equations contains two equations and two variables. For example, consider the system of equations: $x - y = 1, x + y = 5$
- ✓ The easiest way to solve a system of equation is using the elimination method. The elimination method uses the addition property of equality. You can add the same value to each side of an equation.
- ✓ For the first equation above, you can add $x + y$ to the left side and 5 to the right side of the first equation: $x - y + (x + y) = 1 + 5$. Now, if you simplify, you get: $x - y + (x + y) = 1 + 5 \rightarrow 2x = 6 \rightarrow x = 3$. Now, substitute 3 for the x in the first equation: $3 - y = 1$. By solving this equation, $y = 2$

Example:

What is the value of $x + y$ in this system of equations? $\begin{cases} 3x - 4y = -20 \\ -x + 2y = 10 \end{cases}$

Solving Systems of Equations by Elimination: $\begin{array}{r} 3x - 4y = -20 \\ \underline{-x + 2y = 10} \end{array}$ $\Rightarrow$ Multiply the second equation by 3, then add it to the first equation.

$\begin{array}{r} 3x - 4y = -20 \\ \underline{3(-x + 2y = 10)} \end{array} \Rightarrow \begin{array}{r} 3x - 4y = -20 \\ -3x + 6y = 30) \end{array} \Rightarrow 2y = 10 \Rightarrow y = 5$. Now, substitute 5 for y in the first equation and solve for x. $3x - 4(5) = -20 \rightarrow 3x - 20 = -20 \rightarrow x = 0$

Quadratic Equation

Step-by-step guide:

- ✓ Write the equation in the form of: $ax^2 + bx + c = 0$
- ✓ Factorize the quadratic and solve for the variable.
- ✓ Use quadratic formula if you couldn't factorize the quadratic.
- ✓ Quadratic formula: $x = \frac{-b \pm \sqrt{b^2 - 4ac}}{2a}$

Examples:

Find the solutions of each quadratic.

1) $x^2 + 5x - 6 = 0$

Use quadratic formula: $= \frac{-b\pm\sqrt{b^2-4ac}}{2a}$, $a = 1, b = 5$ and $c = -6$

then: $x = \frac{-5\pm\sqrt{5^2-4.1(-6)}}{2(1)}$, $x_1 = \frac{-5+\sqrt{5^2-4\times1(-6)}}{2(1)} = 1$, $x_2 = \frac{-5-\sqrt{5^2-4\times1(-6)}}{2(1)} = -6$

2) $x^2 + 6x + 8 = 0$

Use quadratic formula: $= \frac{-b\pm\sqrt{b^2-4ac}}{2a}$, $a = 1, b = 6$ and $c = 8$

$x = \frac{-6\pm\sqrt{6^2-4.1.8}}{2.1}$, $x_1 = \frac{-6+\sqrt{6^2-4.1.8}}{2.1} = -2$, $x_2 = \frac{-6-\sqrt{6^2-4.1.8}}{2.1} = -4$

Graphing Quadratic Functions

Step-by-step guide:

- ✓ Quadratic functions in vertex form: $y = a(x - h)^2 + k$ where (h, k) is the vertex of the function. The axis of symmetry is $x = h$
- ✓ Quadratic functions in standard form: $y = ax^2 + bx + c$ where $x = -\frac{b}{2a}$ is the value of x in the vertex of the function.
- ✓ To graph a quadratic function, first find the vertex, then substitute some values for x and solve for y.

Example:

Sketch the graph of $\mathbf{y = 3(x + 1)^2 + 2}$.

The vertex of $3(x + 1)^2 + 2$ *is* $(-1,2)$. Substitute zero for x and solve for y. $y = 3(0 + 1)^2 + 2 = 5$. The y Intercept is $(0,5)$.

Now, you can simply graph the quadratic function.

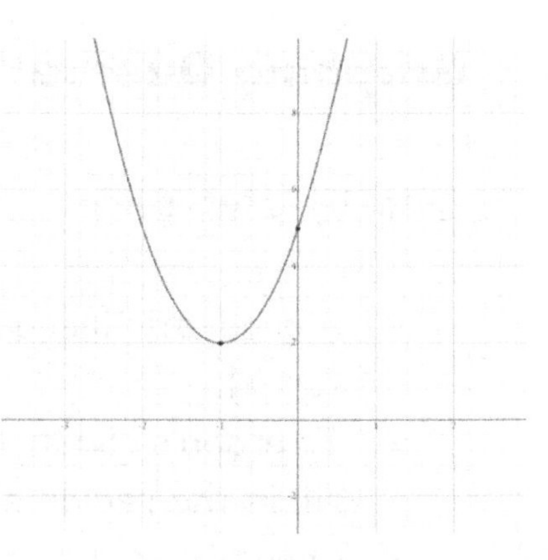

Quadratic Inequalities

Step-by-step guide:

- ✓ A quadratic inequality is one that can be written in one of the following standard forms:

$$ax^2 + bx + c > 0,\ ax^2 + bx + c < 0,\ ax^2 + bx + c \geq 0,\ ax^2 + bx + c \leq 0$$

- ✓ Solving a quadratic inequality is like solving equations. We need to find the solutions.

Examples:

1) ***Solve quadratic inequality.*** $-x^2 - 5x + 6 > 0$

   Factor: $-x^2 - 5x + 6 > 0 \rightarrow -(x-1)(x+6) > 0$

   Multiply both sides by -1: $\left(-(x-1)(x+6)\right)(-1) > 0(-1) \rightarrow (x-1)(x+6) < 0$

   Then the solution could be $-6 < x < 1$ or $-6 > x$ and $x > 1$. Choose a value between -1 and 6 and check. Let's try 0. Then: $-0^2 - 5(0) + 6 > 0 \rightarrow 6 > 0$. This is true! So, the answer is: $-6 < x < 1$

2) ***Solve quadratic inequality.*** $x^2 - 3x - 10 \geq 0$

   Factor: $x^2 - 3x - 10 \geq 0 \rightarrow (x+2)(x-5) \geq 0$. -2 and 5 are the solutions. Now, the solution could be $-2 \leq x \leq 5$ or $-6 \geq x$ and $x \geq 1$. Let's choose zero to check:

   $0^2 - 3(0) - 10 \geq 0 \rightarrow -10 \geq 0$, which is not true. So, $-6 \geq x$ and $x \geq 1$

Graphing Quadratic Inequalities

Step-by-step guide:

- ✓ A quadratic inequality is in the form $y > ax^2 + bx + c$ (or substitute $<, \leq$, or $\geq$ for $>$).
- ✓ To graph a quadratic inequality, start by graphing the quadratic parabola. Then fill in the region either inside or outside of it, depending on the inequality.
- ✓ Choose a testing point and check the solution section.

Example: ***Sketch the graph of*** $y > 2x^2$.

First, graph $y = 2x^2$

Since, the inequality sing is $>$, we need to use dash lines.

Now, choose a testing point inside the parabola. Let's choose $(0,2)$. $y > 2x^2 \rightarrow 2 > 2(0)^2 \rightarrow 2 > 0$

This is true. So, inside the parabola is the solution section.

Adding and Subtracting Complex Numbers

Step-by-step guide:

- ✓ A complex number is expressed in the form $a + bi$, where a and b are real numbers, and i, which is called an imaginary number, is a solution of the equation $x^2 = -1$

- ✓ For adding complex numbers: $(a + bi) + (c + di) = (a + c) + (b + d)i$

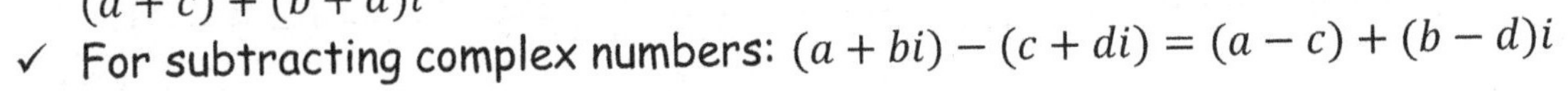

- ✓ For subtracting complex numbers: $(a + bi) - (c + di) = (a - c) + (b - d)i$

Examples:

1) Solve: $-8 + (2i) + (-8 + 6i)$

   Remove parentheses: $-8 + (2i) + (-8 + 6i) \rightarrow -8 + 2i - 8 + 6i$

   Group like terms: $-8 + 2i - 8 + 6i \rightarrow -8 - 8 + 2i + 6i$

   Add similar terms: $-8 - 8 + 2i + 6i = -16 + 8i$

2) Solve: $-2 + (-8 - 7i) - 9$

   Remove parentheses: $-2 + (-8 - 7i) - 9 \rightarrow -2 - 8 - 7i - 9$

   Combine like terms: $2 - 8 - 7i - 9 = -19 - 7i$

Multiplying and Dividing Complex Numbers

Step-by-step guide:

- ✓ Multiplying complex numbers: $(a+bi)+(c+di)=(ac-bd)+(ad+bc)i$
- ✓ Dividing complex numbers: $\frac{a+bi}{c+di}=\frac{a+bi}{c+di}\times\frac{c-di}{c-di}=\frac{ac+bd}{c^2-d^2}+\frac{bc-ad}{c^2-d^2}i$
- ✓ Imaginary number rule: $i^2=-1$

Examples:

1) Solve: $(2-8i)(3-5i)$

   Use the rule: $(a-bi)+(c+di)=(ac-bd)+(ad+bc)i$

   $(2.3-(-8)-5))+(2(-5)+(-8).3)i=-34-34i$

2) Solve: $\frac{2-3i}{2+i}=$

   Use the rule for dividing complex numbers:

$$\frac{a+bi}{c+di}=\frac{a+bi}{c+di}\times\frac{c-di}{c-di}=\frac{ac+bd}{c^2-d^2}+\frac{bc-ad}{c^2-d^2}i\rightarrow$$

$$\frac{2-3i}{2+i}\times\frac{2-i}{2-i}=\frac{2\times(2)+(-3)(1)}{2^2-(i)^2}+\frac{(-3)\times(2)-(2)(1)}{2^2-(-1)}i=\frac{1}{5}-\frac{8}{5}i$$

Rationalizing Imaginary Denominators

Step-by-step guide:

- ✓ Step 1: Find the conjugate (it's the denominator with different sign between the two terms.
- ✓ Step 2: Multiply numerator and denominator by the conjugate.
- ✓ Step 3: Simplify if needed.

Examples:

1) Solve: $\frac{5i}{2-3i}$

Multiply by the conjugate: $\frac{2+3i}{2+3i} \rightarrow \frac{5i(2+3i)}{(2-3i)(2+3i)} = \frac{15i^2+10i}{(2-3i)(2+3i)} = \frac{-15+10i}{(2-3i)(2+3i)}$

Use complex arithmetic rule: $(a+bi)(a-bi) = a^2+b^2$

$(2-3i)(2+3i) = 2^2+(-3)^2 = 13$, Then: $\frac{-15+10i}{(2-3i)(2+3i)} = \frac{-15+10i}{13}$

2) Solve: $\frac{4-9i}{-6i}$

Apply fraction rule: $\frac{4-9i}{-6i} = -\frac{4-9i}{6i}$

Multiply by the conjugate: $\frac{-i}{-i}$. $-\frac{4-9i}{6i} = -\frac{(4-9i)(-i)}{6i(-i)} = -\frac{-9-4i}{6}$

Function Notation

Step-by-step guide:

- ✓ Functions are mathematical operations that assign unique outputs to given inputs.
- ✓ Function notation is the way a function is written. It is meant to be a precise way of giving information about the function without a rather lengthy written explanation.
- ✓ The most popular function notation is $f(x)$ which is read "f of x".

Examples:

1) Evaluate: $w(x) = 3x+1$, find $w(4)$. Substitute x with 4: Then: $w(x) = 3x+1 \rightarrow$ $w(4) = 3(4)+1 \rightarrow w(x) = 12+1 \rightarrow w(x) = 13$

2) Evaluate: $h(n) = n^2 - 10$, find $h(-2)$. Substitute x with -2:

Then: $h(n) = n^2 - 10 \rightarrow h(-2) = (-2)^2 - 10 \rightarrow h(-2) = 4 - 10 \rightarrow h(-2) = -6$

Adding and Subtracting Functions

Step-by-step guide:

- ✓ Just like we can add and subtract numbers, we can add and subtract functions. For example, if we had functions f and g, we could create two new functions:
- ✓ f + g and f - g.

Examples:

1) $f(x) = 2x + 4, g(x) = x + 3$, Find: $(f - g)(1)$

   $(f - g)(x) = f(x) - g(x)$, then: $(f - g)(x) = 2x + 4 - (x + 3)$

   $= 2x + 4 - x - 3 = x + 1$

   Substitute x with 1: $(f - g)(1) = 1 + 1 = 2$

2) $g(a) = 2a - 1, f(a) = -a - 4$, Find: $(g + f)(-1)$

   $(g + f)(a) = g(a) + f(a)$, Then: $(g + f)(a) = 2a - 1 - a - 4 = a - 5$

   Substitute a with -1: $(g + f)(a) = a - 5 = -1 - 5 = -6$

Multiplying and Dividing Functions

Step-by-step guide:

- ✓ Just like we can multiply and divide numbers, we can multiply and divide functions. For example, if we had functions f and g, we could create two new functions: f × g, and $\frac{f}{g}$.

Examples:

1) $g(x) = -x - 2, f(x) = 2x + 1$, Find: $(g.f)(2)$

   $(g.f)(x) = g(x).f(x) = (-x - 2)(2x + 1) = -2x^2 - x - 4x - 2 = -2x^2 - 5x - 2$

   Substitute x with 2:

   $(g.f)(x) = -2x^2 - 5x - 2 = -2(2)^2 - 5(2) - 2 = -8 - 10 - 2 = -20$

2) $f(x) = x + 4, h(x) = 5x - 2$, Find: $\left(\frac{f}{h}\right)(-1)$

$\left(\frac{f}{h}\right)(x) = \frac{f(x)}{h(x)} = \frac{x+4}{5x-2}$

Substitute x with -1: $\left(\frac{f}{h}\right)(x) = \frac{x+4}{5x-2} = \frac{(-1)+4}{5(-1)-2} = \frac{3}{-7} = -\frac{3}{7}$

Composition of Functions

Step-by-step guide:

- ✓ The term "composition of functions" (or "composite function") refers to the combining together of two or more functions in a manner where the output from one function becomes the input for the next function.
- ✓ The notation used for composition is:$(f\ o\ g)(x) = f(g(x))$

Examples:

1) *Using* $\mathrm{f(x) = x + 2}$ *and* $\mathrm{g(x) = 4x}$, *find:* $f(g(1))$

$(f\ o\ g)(x) = f(g(x))$

Then: $(f\ o\ g)(x) = f\big(g(x)\big) = f(4x) = 4x + 2$

Substitute x with 1: $(f\ o\ g)(1) = 4 + 2 = 6$

2) *Using* $\mathrm{f(x) = 5x + 4}$ *and* $\mathrm{g(x) = x - 3}$, *find:* $\boldsymbol{g(f(3))}$

$(f\ o\ g)(x) = f(g(x))$

Then: $(g\ o\ f)(x) = g\big(\mathrm{f(x)}\big) = \mathrm{g(5x + 4)}$, *now substitute* x *in* $\mathrm{g(x)}$ *by* $\mathrm{5x + 4}$. *Then:*

$g(\mathrm{5x} + 4) = (5x + 4) - 3 = 5x + 4 - 3 = 5x + 1$

Substitute x with 3: $(g\ o\ f)(x) = g\big(f(x)\big) = 5x + 1 = 5(3) + 1 = 15 = 1 = 16$

Trig Ratios of General Angles

Step-by-step guide:

✓ Learn common trigonometric functions:

θ	$0°$	$30°$	$45°$	$60°$	$90°$
$\sin\theta$	0	$\frac{1}{2}$	$\frac{\sqrt{2}}{2}$	$\frac{\sqrt{3}}{2}$	1
$\cos\theta$	1	$\frac{\sqrt{3}}{2}$	$\frac{\sqrt{2}}{2}$	$\frac{1}{2}$	0
$\tan\theta$	0	$\frac{\sqrt{3}}{3}$	1	$\sqrt{3}$	Undefined

Examples:

Find each trigonometric function.

1) $sin - 120°$. Use the following property: $sin(-x) = -sin(x)$

$sin - 120° = -sin\ 120°$. $sin\ 120° = \frac{\sqrt{3}}{2}$, then: $sin - 120° = -\frac{\sqrt{3}}{2}$

2) $cos\ 150°$

Recall that $cos\ 150° = -cos\ 30°$. Then: $cos\ 150° = -cos\ 30° = -\frac{\sqrt{3}}{2}$

Angles and Angle Measure

Step-by-step guide:

✓ To convert degrees to radians, use this formula: $\boldsymbol{Radians = Degrees \times \frac{\pi}{180}}$

✓ To convert radians to degrees, use this formula: $\boldsymbol{Degrees = Radians \times \frac{180}{\pi}}$

Examples:

1) Convert 150 degrees to radians.

Use this formula: $Radians = Degrees \times \frac{\pi}{180}$

$$Radians = 150 \times \frac{\pi}{180} = \frac{150\pi}{180} = \frac{5\pi}{6}$$

2) Convert $\frac{2\pi}{3}$ to degrees.

Use this formula: $Degrees = Radians \times \frac{180}{\pi}$

$$Degrees = \frac{2\pi}{3} \times \frac{180}{\pi} = \frac{360\pi}{3\pi} = 120°$$

Evaluating Trigonometric Function

Step-by-step guide:

- ✓ Step 1: Draw the terminal side of the angle.
- ✓ Step 2: Find reference angle. (It is the smallest angle that you can make from the terminal side of an angle with the x-axis.)
- ✓ Step 3: Find the trigonometric function of the reference angle.

Examples:

1) ***Find the exact value of trigonometric function.*** $\cos 225°$

Write $\cos(225°)$ as $\cos(180° + 45°)$. Recall that $\cos 180° = -1, \cos 45° = \frac{\sqrt{2}}{2}$

225° is in the third quadrant and cosine is negative in the quadrant 3. T*he reference angle of* 225° is 45°. Therefore, $\cos 225° = -\frac{\sqrt{2}}{2}$

2) ***Find the exact value of trigonometric function.*** $\tan \frac{7\pi}{6}$

Rewrite the angles for $\tan \frac{7\pi}{6}$:

$\tan \frac{7\pi}{6} = \tan\left(\frac{6\pi+\pi}{6}\right) = \tan(\pi + \frac{1}{6}\pi)$

Use the periodicity of tan: $\tan(x + \pi . k) = \tan(x)$

$\tan\left(\pi + \frac{1}{6}\pi\right) = \tan\left(\frac{1}{6}\pi\right) = \frac{\sqrt{3}}{3}$

Day 7 Practices

Solve each system of equations.

1) $-5x + y = -3$ $\quad x =$ ____
$3x - 8y = 24$ $\quad y =$ ____

2) $3x - 2y = 2$ $\quad x =$ ____
$5x - 5y = 10$ $\quad y =$ ____

3) $8x + 14y = 4$ $\quad x =$ ____
$-6x - 7y = -10$ $\quad y =$ ____

4) $10x + 7y = 1$ $\quad x =$ ____
$-5x - 7y = 24$ $\quad y =$ ____

Factor each expression.

5) $x^2 - 5x + 4 =$
6) $x^2 + 6x + 8 =$
7) $x^2 + x - 12 =$
8) $x^2 - 7x + 10 =$
9) $x^2 - 4x - 12 =$
10) $2x^2 - 3x - 2 =$

Solve each quadratic inequality.

11) $x^2 + 4x - 5 > 0$
12) $x^2 - 2x - 3 \geq 0$
13) $x^2 - 1 < 0$
14) $17x^2 + 15x - 2 \geq 0$
15) $4x^2 + 20x - 11 < 0$
16) $12x^2 + 10x - 12 > 0$

Simplify.

17) $(-3 + 6i) - (-9 - i) =$
18) $(-5 + 15i) - (-3 + 3i) =$
19) $(-14 + i) - (-12 - 11i) =$
20) $(-18 - 3i) + (11 + 5i) =$
21) $(-11 - 9i) - (-9 - 3i) =$
22) $-8 + (2i) + (-8 + 6i) =$

Simplify.

23) $(-2 - i)(4 + i) =$
24) $(2 - 2i)^2 =$
25) $(4 - 3i)(6 - 6i) =$
26) $\frac{-1+5i}{-8-7i} =$
27) $\frac{-2-9i}{-2+7i} =$
28) $\frac{4+i}{2-5i} =$

Simplify.

29) $\frac{-2}{-2i} =$
30) $\frac{-1}{-9i} =$
31) $\frac{-8}{-5i} =$
32) $\frac{-6-i}{-1+6i} =$
33) $\frac{-9-3i}{-3+3i} =$
34) $\frac{4i+1}{-1+3i} =$

Evaluate each function.

35) $f(x) = x - 2$, find $f(1)$
36) $g(x) = 2x + 3$, find $g(2)$
37) $h(x) = x + 8$, find $h(5)$
38) $h(n) = n^2 + 4$, find $h(-4)$
39) $h(n) = n^2 - 10$, find $h(5)$
40) $h(n) = -2n^2 - 6n$, find $h(2)$

Perform the indicated operation.

41) $g(a) = -3a - 3$
$f(a) = a^2 + 5$
Find $(g - f)(a)$

42) $g(t) = 2t + 5$
$f(t) = -t^2 + 5$
Find $(g + f)(t)$

Perform the indicated operation.

43) $g(x) = -x - 2$
$f(x) = 2x + 1$
Find $(g.f)(2)$

44) $f(x) = 3x$
$h(x) = -2x + 5$
Find $(f.h)(-1)$

Using $f(x) = 5x + 4$ ***and*** $g(x) = x - 3$***, find:***

45) $g(f(-3)) =$
46) $g(f(4)) =$
47) $f(g(6)) =$
48) $f(f(8)) =$

Find the exact value of each trigonometric function. Some may be undefined.

49) $sec\,\pi =$
50) $tan -\frac{3\pi}{2} =$
51) $cos\,\frac{11\pi}{6} =$
52) $cot\,\frac{5\pi}{3} =$
53) $sec -\frac{3\pi}{4} =$
54) $sec\,\frac{\pi}{3} =$

Convert each degree measure into radians and each radian measure into degrees.

55) $420° =$ ____
56) $300° =$ ____
57) $-60° =$ ____
58) $-\frac{16\pi}{3} =$
59) $-\frac{3\pi}{5} =$
60) $\frac{11\pi}{6} =$

Use the given point on the terminal side of angle θ to find the value of the trigonometric function indicated.

61) $sin\theta;\ (-6, 4)$
62) $cos\theta;\ (2, -2)$
63) $cot\theta;\ (-7, \sqrt{15})$
64) $cos\theta;\ (-5, -12)$
65) $sin\theta;\ (-\sqrt{7}, 3)$
66) $tan\theta;\ (-11, -2)$

Answers

1) $x = 0, y = -3$
2) $x = -2, y = -4$
3) $x = 4, y = -2$
4) $x = 5, y = -7$
5) $(x-4)(x-1)$
6) $(x+4)(x+2)$
7) $(x-3)(x+4)$
8) $(x-5)(x-2)$
9) $(x+2)(x-6)$
10) $(2x+1)(x-2)$
11) $x < -5$ or $x > 1$
12) $x \le -1$ or $x \ge 3$
13) $-1 < x < 1$
14) $x \le -1$ or $x \ge \frac{2}{17}$
15) $-\frac{11}{2} < x < \frac{1}{2}$
16) $x < -\frac{3}{2}$ or $x > \frac{2}{3}$
17) $6 + 7i$
18) $-2 + 12i$
19) $-2 + 12i$
20) $-7 + 2i$
21) $-2 - 6i$
22) $-16 + 8i$
23) $-7 - 6i$
24) $-8i$
25) $6 - 42i$
26) $-\frac{27}{113} - \frac{47}{113}i$
27) $-\frac{59}{53} + \frac{32}{53}i$
28) $\frac{3}{29} + \frac{22}{29}i$
29) $-i$
30) $-\frac{1}{9}i$
31) $\frac{-8}{5}i$
32) i
33) $1 + 2i$
34) $\frac{11}{10} - \frac{7}{10}i$
35) -1
36) 7
37) 13
38) 20
39) 15
40) -20
41) $-a^2 - 3a - 8$
42) $-t^2 + 2t + 10$
43) -20
44) -21
45) -14
46) 21
47) 19
48) 224
49) -1
50) Undefined
51) $\frac{\sqrt{3}}{2}$
52) $-\frac{\sqrt{3}}{3}$
53) $-\sqrt{2}$
54) 2
55) $\frac{7\pi}{3}$
56) $\frac{5\pi}{3}$
57) $-\frac{\pi}{3}$
58) $-960°$
59) $-108°$
60) $330°$
61) $\frac{2\sqrt{13}}{13}$
62) $\sqrt{2}$
63) $-\frac{7\sqrt{15}}{15}$
64) $-\frac{5}{13}$
65) $\frac{3}{4}$
66) $\frac{2}{11}$

Next Generation Accuplacer Test Review

The Next-Generation ACCUPLACER test is an assessment system for measuring students' readiness for college courses in reading, writing, and mathematics. The test is a multiple–choice format and is used to precisely placing you at the correct level of introductory classes.

The Next-Generation ACCUPLACER uses the computer–adaptive technology and the questions you see are based on your skill level. Your response to each question drives the difficulty level of the next question.

There are five sub sections on the Accuplacer test:

- ✓ Arithmetic (20 questions)
- ✓ Quantitative Reasoning, Algebra, And Statistics (QAS) (20 questions)
- ✓ Advanced Algebra and Functions (20 questions)
- ✓ Reading (20 questions)
- ✓ Writing (25 questions)

Accuplacer does NOT permit the use of personal calculators on the Math portion of placement test. Accuplacer expects students to be able to answer certain questions without the assistance of a calculator. Therefore, they provide an onscreen calculator for students to use on some questions.

In this section, there are two complete Accuplacer Mathematics Tests. Take these tests to see what score you'll be able to receive on a real Accuplacer test.

Good luck!

Time to refine your Math skill with a practice test

Take an Accuplacer Next Generation test to simulate the test day experience. After you've finished, score your test using the answer keys.

Before You Start

- You'll need a pencil to take the test.
- For each question, there are four possible answers. Choose which one is best.
- It's okay to guess. There is no penalty for wrong answers.
- After you've finished the test, review the answer key to see where you went wrong.

Good Luck!

Mathematics is like love; a simple idea, but it can get complicated.

Next Generation Accuplacer Mathematics Practice Test 1

2019 - 2020

Section 1: Arithmetic

(No Calculator)

20 questions

Total time for this section: No time limit.

You may NOT use a calculator on this Section.

(On a real Accuplacer test, there is **an onscreen calculator to use on some questions.)**

1) Which of the following inequalities is true?
   A. $\frac{3}{4} < \frac{17}{24}$
   B. $\frac{2}{3} < \frac{7}{9}$
   C. $\frac{3}{8} < \frac{9}{25}$
   D. $\frac{11}{21} < \frac{4}{7}$

2) What is the value of 4.56×7.8?
   A. 35.568
   B. 36.08
   C. 355.68
   D. 360.80

3) 12% of what number is equal to 72?
   A. 11.25
   B. 112.50
   C. 400
   D. 600

4) Which of the following is greater than $\frac{13}{8}$?
   A. $\frac{1}{2}$
   B. $\frac{5}{2}$
   C. $\frac{3}{4}$
   D. 1

5) A taxi driver earns $\$8$ per 1-hour work. If he works 10 hours a day and in 1 hour he uses 2-liters petrol with price $\$1$ for 1-liter. How much money does he earn in one day?
   A. $\$90$
   B. $\$88$
   C. $\$60$
   D. $\$50$

6) Which of the following is closest to 6.03?
   A. 7
   B. 6.5
   C. 6
   D. 6.4

7) The price of a sofa is decreased by 20% to \$476. What was its original price?
   A. \$480
   B. \$520
   C. \$595
   D. \$600

8) If 45% of a class are girls, and 25% of girls play tennis, what percent of the class play tennis?
   A. 11%
   B. 15%
   C. 20%
   D. 40%

9) If 60% of A is 30% of B, then B is what percent of A?
   A. 2%
   B. 20%
   C. 100%
   D. 200%

10) The price of a car was \$20,000 in 2014, \$16,000 in 2015 and \$12,800 in 2016. What is the rate of depreciation of the price of car per year?
   A. 15%
   A. 20%
   C. 25%
   D. 30%

11) A bank is offering 3.5% simple interest on a savings account. If you deposit \$9,000, how much interest will you earn in five years?
   A. \$360
   B. \$720
   C. \$1,575
   D. \$3,600

12) Which of the following expressions has the same value as $\frac{5}{4} \times \frac{6}{2}$?
   A. $\frac{6\times3}{4}$
   B. $\frac{6\times2}{4}$
   C. $\frac{5\times6}{4}$
   D. $\frac{5\times3}{4}$

13) 35 is What percent of 20?

A. 20%
B. 25%
C. 175%
D. 190%

14) What is the value of $3.85 + 0.045 + 0.1365$?

A. 2.2565
B. 4.0315
C. 4.215
D. 4.4265

15) How long does a 420–miles trip take moving at 65 miles per hour (mph)?

A. $4\ hours$
B. $4\ hours\ and\ 24\ minutes$
C. $6\ hours\ and\ 24\ minutes$
D. $8\ hours\ and\ 30\ minutes$

16) Which of the following lists shows the fractions in order from least to greatest?

$$\frac{5}{4}, \frac{2}{7}, \frac{3}{8}, \frac{5}{11}$$

A. $\frac{3}{8}, \frac{2}{7}, \frac{5}{4}, \frac{5}{11}$
B. $\frac{2}{7}, \frac{5}{11}, \frac{3}{8}, \frac{5}{4}$
C. $\frac{2}{7}, \frac{3}{8}, \frac{5}{11}, \frac{5}{4}$
D. $\frac{3}{8}, \frac{2}{7}, \frac{5}{11}, \frac{5}{4}$

17) $\frac{4}{5} - \frac{2}{5} = ?$

A. 0.3
B. 0.35
C. 0.4
D. 0.45

18) $\frac{(8+6)^2}{2} + 6 = ?$

A. 110
B. 104
C. 90
D. 14

19) A rope weighs 600 grams per meter of length. What is the weight in kilograms of 14.2 meters of this rope? ($1\ kilograms = 1000\ grams$)

A. 0.0852
B. 0.852
C. 8.52
D. 85.20

20) When 78 is divided by 5, the remainder is the same as when 45 is divided by

A. 2
B. 4
C. 5
D. 7

STOP: This is the End of Section 1 of test 1.

Next Generation Accuplacer Mathematics Practice Test 1

2019 - 2020

Section 2: Quantitative Reasoning, Algebra, And Statistics

(No Calculator)

20 questions

Total time for this section: No time limit.

You may NOT use a calculator on this Section.

(*On a real Accuplacer test, there is* an onscreen calculator to use on some questions.)

1) When a number is subtracted from 28 and the difference is divided by that number, the result is 3. What is the value of the number?
   A. 2
   B. 4
   C. 7
   D. 12

2) An angle is equal to one fifth of its supplement. What is the measure of that angle?
   A. 30
   B. 34
   C. 36
   D. 45

3) John traveled 150 km in 5 hours and Alice traveled 180 km in 4 hours. What is the ratio of the average speed of John to average speed of Alice?
   A. $3:2$
   B. $2:5$
   C. $2:3$
   D. $5:6$

4) A taxi driver earns \$7 per 1-hour work. If he works 10 hours a day and in 1 hour he uses 2-liters petrol with price \$1 for 1-liter. How much money does he earn in one day?
   A. \$90
   B. \$88
   C. \$50
   D. \$40

5) Right triangle ABC has two legs of lengths 6 cm (AB) and 8 cm (AC). What is the length of the third side (BC)?
   A. 6 cm
   B. 8 cm
   C. 10 cm
   D. 14 cm

6) The area of a circle is less than 49π. Which of the following can be the circumference of the circle?
   A. 12π
   B. 14π
   C. 24π
   D. 32π

7) The width of a box is one third of its length. The height of the box is one third of its width. If the length of the box is $36\ cm$, what is the volume of the box?
   A. $81\ cm^3$
   B. $162\ cm^3$
   C. $C.243\ cm^3$
   D. $1{,}728\ cm^3$

8) How many possible outfit combinations come from six shirts, four slacks, and five ties?
   A. 15
   B. 18
   C. 30
   D. 120

9) If the area of trapezoid is $100\ cm$, what is the perimeter of the trapezoid?
   A. $12\ cm$
   B. $32\ cm$
   C. $35\ cm$
   D. $55\ cm$

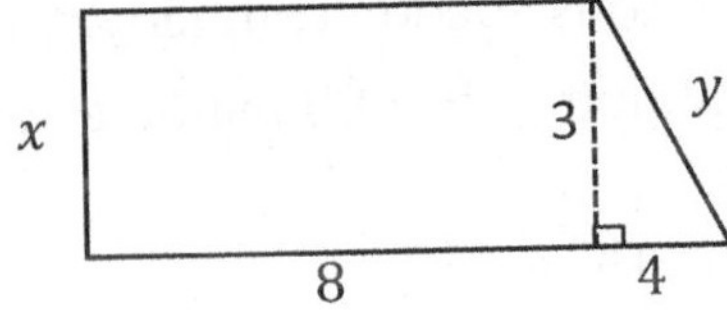

10) In the xy-plane, the point $(4,3)$ and $(3,2)$ are on line A. Which of the following points could also be on line A?
   A. $(-1,2)$
   B. $(5,7)$
   C. $(3,4)$
   D. $(-1,-2)$

11) Two third of 15 is equal to $\frac{2}{5}$ of what number?
   A. 12
   B. 20
   C. 25
   D. 60

12) A boat sails 120 miles south and then 50 miles east. How far is the boat from its start point?
   A. $45\ miles$
   B. $130\ miles$
   C. $160\ miles$
   D. $170\ miles$

13) The ratio of boys and girls in a class is $4:7$. If there are 55 students in the class, how many more boys should be enrolled to make the ratio $1:1$?

A. 8
B. 10
C. 15
D. 18

14) The score of Emma was half as that of Ava and the score of Mia was twice that of Ava. If the score of Mia was 40, what is the score of Emma?

A. 5
B. 10
C. 20
D. 30

15) A construction company is building a wall. The company can build $40\ cm$ of the wall per minute. After 50 minutes construction, $\frac{2}{3}$ of the wall is completed. How high is the wall?

A. $10\ m$
B. $15\ m$
C. $30\ m$
D. $35\ m$

16) The average of five consecutive numbers is 40. What is the smallest number?

A. 40
B. 38
C. 34
D. 12

17) The average weight of 18 girls in a class is 55 kg and the average weight of 32 boys in the same class is 62 kg. What is the average weight of all the 50 students in that class?

A. 60
B. 61.28
C. 61.68
D. 62.90

18) What is the median of these numbers? $3, 9, 12, 8, 15, 19, 5$

A. 8
B. 9
C. 13
D. 15

19) The surface area of a cylinder is $150\pi\ cm^2$. If its height is $10\ cm$, what is the radius of the cylinder?

A. $13\ cm$
B. $11\ cm$
C. $15\ cm$
D. $5\ cm$

20) In 1999, the average worker's income increased \$2,000 per year starting from \$25,000 annual salary. Which equation represents income greater than average? (I = income, x = number of years after 1999)

A. $I > 2{,}000\ x + 25{,}000$
B. $I > -2{,}000\ x + 25{,}000$
C. $I < -2{,}000\ x + 25{,}000$
D. $I < 2{,}000\ x - 25{,}000$

STOP: This is the End of Section 2 of test 1.

Next Generation Accuplacer Mathematics Practice Test 1

2019 - 2020

Section 3: Advanced Algebra and Functions

(Calculator)

20 questions

Total time for this section: No time limit.

You may use a calculator on this Section.

(On a real Accuplacer test, there is an onscreen calculator to use on some questions.)

1) If $x + y = 0,\ 4x - 2y = 24$, which of the following ordered pairs (x, y) satisfies both equations?
   A. $(4, 3)$
   B. $(5, 4)$
   C. $(4, -4)$
   D. $(4, -6)$

2) If $f(x) = 3x + 4(x + 1) + 2$ then $f(3x) =?$
   A. $21x + 6$
   B. $16x - 6$
   C. $25x + 4$
   D. $12x + 3$

3) A line in the xy-plane passes through origin and has a slope of $\frac{1}{3}$. Which of the following points lies on the line?
   A. $(2,1)$
   B. $(4,1)$
   C. $(9,3)$
   D. $(6,3)$

4) Which of the following is equivalent to $(3n^2 + 4n + 6) - (2n^2 - 5)$?
   A. $n + 4n^2$
   B. $n^2 - 3$
   C. $n^2 + 4n + 11$
   D. $n + 2$

5) If $(ax + 4)(bx + 3) = 10x^2 + cx + 12$ for all values of x and $a + b = 7$, what are the two possible values for c?
   A. $22, 21$
   B. $20, 22$
   C. $23, 26$
   D. $24, 23$

6) If $x \neq -4$ and $x \neq 6$, which of the following is equivalent to $\frac{1}{\frac{1}{x-6}+\frac{1}{x+4}}$?

A. $\frac{(x-6)(x+4)}{(x-6)+(x+4)}$
B. $\frac{(x+4)+(x-6)}{(x+4)(x-6)}$
C. $\frac{(x+4)(x-6)}{(x+4)-(x+6)}$
D. $\frac{(x+4)+(x-6)}{(x+4)-(x-6)}$

$$y < a - x \, , y > x + b$$

7) In the xy-plane, if $(0, 0)$ is a solution to the system of inequalities above, which of the following relationships between a and b must be true?
A. $a < b$
B. $a > b$
C. $a = b$
D. $a = \ b + \ a$

8) Which of the following points lies on the line that goes through the points $(2, 4)$ and $(4, 5)$?
A. $(9,9)$
B. $(9,6)$
C. $(6,9)$
D. $(6,6)$

9) Calculate $f(4)$ for the following function f.

$$f(x) = x^2 - 3x$$

A. 0

B. 4

C. 12

D. 20

10) John buys a pepper plant that is 6 inches tall. With regular watering the plant grows 4 inches a year. Writing John's plant's height as a function of time, what does the y –intercept represent?
A. The y –intercept represents the rate of grows of the plant which is 4 inches
B. The y –intercept represents the starting height of 6 inches
C. The y –intercept represents the rate of growth of plant which is 4 inches per year
D. There is no y –intercept

11) If $\frac{3}{x} = \frac{12}{x-9}$ what is the value of $\frac{x}{6}$?

A. -2
B. 2
C. $-\frac{1}{2}$
D. $\frac{1}{2}$

12) Which of the following is an equation of a circle in the xy-plane with center $(0, 4)$ and a radius with endpoint $(\frac{5}{3}, 6)$?

A. $(x+1)^2 + (y-4)^2 = \frac{61}{9}$
B. $2x^2 + (y+4)^2 = \frac{61}{9}$
C. $(x-2)^2 + (y-4)^2 = \frac{61}{9}$
D. $x^2 + (y-4)^2 = \frac{61}{9}$

13) Given a right triangle ΔABC whose $n\angle B = 90^{\circ}, \sin C = \frac{2}{3}$, find $\cos A$?

A. 1
B. $\frac{1}{2}$
C. $\frac{2}{3}$
D. $\frac{3}{2}$

14) What is the equation of the following graph?

A. $x^2 + 6x + 5$
B. $x^2 + 2x + 4$
C. $2x^2 - 4x + 4$
D. $2x^2 + 4x + 2$

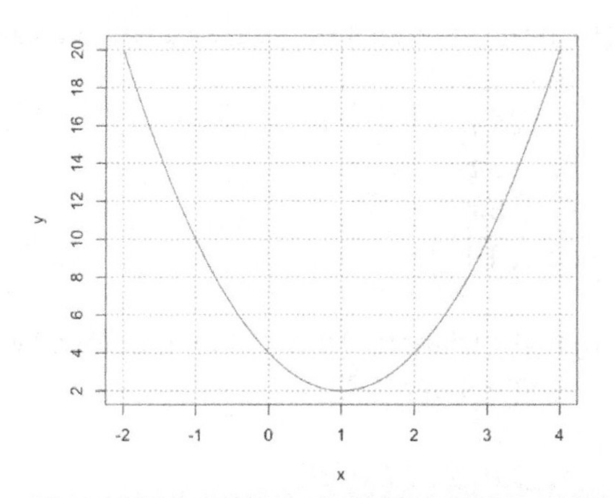

15) In the xy –plane, the line determined by the points $(6, m)$ and $(m, 12)$ passes through the origin. Which of the following could be the value of m?

A. $\sqrt{6}$
B. 12
C. $6\sqrt{2}$
D. 9

16) A function $g(3) = 5$ and $g(6) = 4$. A function $f(5) = 2$ and $f(4) = 7$. What is the value of $f(g(6))$?

A. 5
B. 7
C. 8
D. 9

17) What is the area of the following equilateral triangle if the side $AB = 8\ cm$?

A. $16\sqrt{3}\ cm^2$
B. $8\sqrt{3}\ cm^2$
C. $\sqrt{3}\ cm^2$
D. $8\ cm^2$

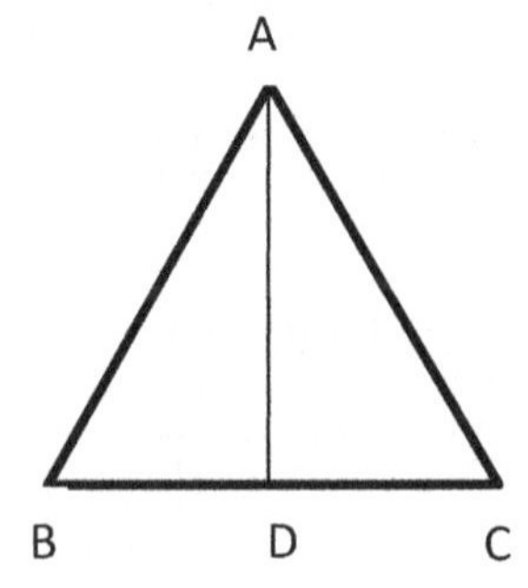

18) A function $g(x)$ satisfies $g(4) = 5$ and $g(7) = 8$. A function $f(x)$ satisfies $f(5) = 18$ and $f(8) = 35$. What is the value of $f(g(7))$?

A. 12
B. 22
C. 35
D. 42

$$(x + 2)^2 + (y - 4)^2 = 16$$

19) In the standard (x, y) coordinate system plane, what is the area of the circle with the above equation?

A. 24π
B. 18π
C. 16π
D. $\sqrt{10}$

20) Right triangle ABC is shown below. Which of the following is true for all possible values of angle A and B?

A. $tan\ A = tan\ B$
B. $sin\ A = cos\ B$
C. $tan^2 A = tan^2 B$
D. $tan\ A = 1$

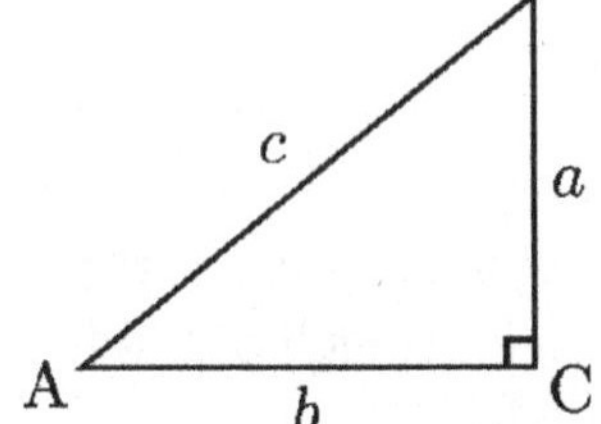

STOP: This is the End of Test 1.

Next Generation Accuplacer Mathematics Practice Test 2

2019 - 2020

Section 1: Arithmetic

(No Calculator)

20 questions

Total time for this section: No time limit.

You may NOT use a calculator on this Section.

(On a real Accuplacer test, there is **an onscreen calculator to use on some questions.)**

1) What is 5 percent of 560?
   A. 20
   B. 28
   C. 30
   D. 40

2) In two successive years, the population of a town is increased by 10% and 20%. What percent of the population is increased after two years?
   A. 68%
   B. 35%
   C. 32%
   D. 28%

3) For what price is 20 percent off the same as $75 off?
   A. $200
   B. $300
   C. $350
   D. $375

4) Last week 25,000 fans attended a football match. This week three times as many bought tickets, but one sixth of them cancelled their tickets. How many are attending this week?
   A. 48,000
   B. 54,000
   C. 62,500
   D. 72,000

5) $\frac{1\frac{5}{4}+\frac{1}{3}}{2\frac{1}{2}-\frac{15}{8}}$ is approximately equal to.
   A. 4.133
   B. 4.6
   C. 5.67
   D. 6.33
   E. 6.67

6) What is $12{,}181 + 8{,}951$?
   A. 21,132
   B. 21,872
   C. 22,776
   D. 23,771

7) Bob deposits 20% of $150 into a savings account, what is the amount of his deposit?
   A. $10
   B. $16
   C. $20
   D. $30

8) If 150% of a number is 75, then what is the 80% of that number?
   A. 40
   B. 50
   C. 70
   D. 85

9) What is the remainder when 754 is divided by 7?
   A. 2
   B. 3
   C. 5
   D. 6

10) Which of the following fractions is less than $\frac{3}{2}$?
   A. 1.3
   B. $\frac{5}{2}$
   C. 3.01
   D. 2.7

11) Mr. Jones saves $2,500 out of his monthly family income of $65,000. What fractional part of his income does he save?
   A. $\frac{1}{26}$
   B. $\frac{1}{11}$
   C. $\frac{3}{25}$
   D. $\frac{2}{15}$

12) 15% of what number is equal to 75?
   A. 8.64
   B. 36
   C. 300
   D. 500

13) 44 students took an exam and 11 of them failed. What percent of the students passed the exam?

A. 20%
B. 40%
C. 60%
D. 75%

14) A bank is offering 3.5% simple interest on a savings account. If you deposit $13,000, how much interest will you earn in two years?

A. $420
B. $910
C. $4200
D. $8,400

15) What is 0.6749 rounded to the nearest hundredth?

A. 0.67
B. 0.675
C. 0.68
D. 0.684

16) If a gas tank can hold 28 gallons, how many gallons does it contain when it is $\frac{3}{4}$ full?

A. 27
B. 25
C. 24
D. 21

17) Jack earns $616 for his first 44 hours of work in a week and is then paid 1.5 times his regular hourly rate for any additional hours. This week, Jack needs $826 to pay his rent, bills and other expenses. How many hours must he work to make enough money in this week?

A. 40
B. 48
C. 53
D. 54

18) What is 420.756 rounded to the nearest hundredth?

A. 420
B. 420.75
C. 420.76
D. 421

$$\frac{5}{8}, 0.625\%, 0.0625$$

19) Which of the following correctly orders the values above from least to greatest?

A. $0.625\%, \frac{5}{8}, 0.0625$
B. $0.0625, 0.625\%, \frac{5}{8}$
C. $\frac{5}{8}, 0.625\%, 0.0625$
D. $\frac{5}{8}, 0.0625, 0.625\%$

20) Which of the following is equivalent to $\frac{3}{5}$?

A. 0.06
B. 0.25
C. 0.60
D. 1.4

STOP: This is the End of Section 1 of test 2.

Next Generation Accuplacer Mathematics Practice Test 2

2019 - 2020

Section 2: Quantitative Reasoning, Algebra, And Statistics

(No Calculator)

20 questions

Total time for this section: No time limit.

You may NOT use a calculator on this Section.

(*On a real Accuplacer test, there is* an onscreen calculator to use on some questions.)

1) In a stadium the ratio of home fans to visiting fans in a crowd is $5:7$. Which of the following could be the total number of fans in the stadium?
   A. $12{,}324$
   B. $42{,}326$
   C. $44{,}566$
   D. $66{,}812$

2) Which of the following points lies on the line $x + 2y = 4$?
   A. $(-2, 3)$
   B. $(1, 2)$
   C. $(-1, 3)$
   D. $(-3, 4)$

3) The perimeter of a rectangular yard is 72 meters. What is its length if its width is twice its length?
   A. $12\ meters$
   B. $18\ meters$
   C. $20\ meters$
   D. $24\ meters$

4) The mean of 50 test scores was calculated as 86. But, it turned out that one of the scores was misread as 91 but it was 66. What is the correct mean of the test scores?
   A. 84
   B. 85
   C. 85.5
   D. 86.5

5) If $3x + y = 25$ and $x - z = 14$, what is the value of x?
   A. 0
   B. 5
   C. 10
   D. It cannot be determined from the information given

6) A swimming pool holds 2,500 cubic feet of water. The swimming pool is 25 feet long and 10 feet wide. How deep is the swimming pool?
   A. 2
   B. 4
   C. 6
   D. 10

7) What is the area of a square whose diagonal is 6?
   A. 6
   B. 18
   C. 32
   D. 64

8) Anita's trick–or–treat bag contains 15 pieces of chocolate, 10 suckers, 10 pieces of gum, 25 pieces of licorice. If she randomly pulls a piece of candy from her bag, what is the probability of her pulling out a piece of sucker?
   A. $\frac{1}{3}$
   B. $\frac{1}{6}$
   C. $\frac{1}{8}$
   D. $\frac{1}{12}$

9) The average of 6 numbers is 14. The average of 4 of those numbers is 10. What is the average of the other two numbers?
   A. 10
   B. 12
   C. 14
   D. 22

10) The perimeter of the trapezoid below is $46\ cm$. What is its area?

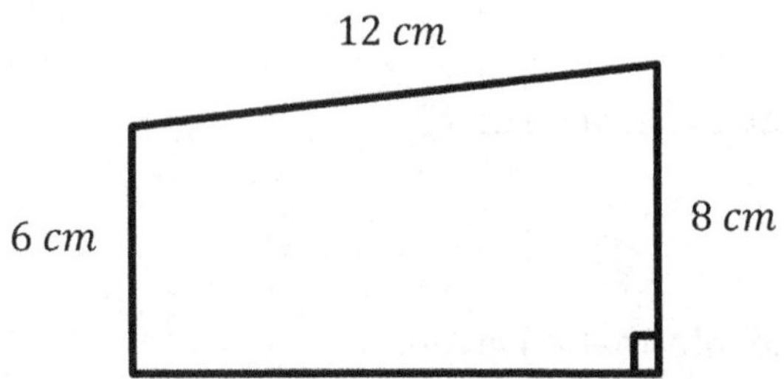

   A. $576\ cm^2$
   B. $140\ cm^2$
   C. $48\ cm^2$
   D. $24\ cm^2$

11) A football team had $20,000 to spend on supplies. The team spent $15,000 on new balls. New sport shoes cost $120 each. Which of the following inequalities represent the number of new shoes the team can purchase?
   A. $120x + 15{,}000 \leq 20{,}000$
   B. $120x + 15{,}000 \geq 20{,}000$
   C. $15{,}000x + 120 \leq 20{,}000$
   D. $15{,}000x + 120 \geq 20{,}000$

12) A card is drawn at random from a standard 52–card deck, what is the probability that the card is of Hearts? (The deck includes 13 of each suit clubs, diamonds, hearts, and spades)

A. $\frac{1}{3}$
B. $\frac{1}{4}$
C. $\frac{1}{6}$
D. $\frac{1}{52}$

13) The average of five numbers is 30. If a sixth number that is greater than 42 is added, then, which of the following could be the new average? (Select one or more answer choices)

A. 25
B. 26
C. 32
D. 38

14) The length of a rectangle is 3 meters greater than 4 times its width. The perimeter of the rectangle is 36 meters. What is the area of the rectangle in meters?

A. 35
B. 45
C. 55
D. 65

15) The ratio of boys and girls in a class is $3:8$. If there are 66 students in the class, how many more boys should be enrolled to make the ratio $1:1$?

A. 6
B. 18
C. 20
D. 30

16) What is the value of x in the following equation?

$$\frac{2}{3}x + \frac{1}{6} = \frac{1}{2}$$

A. 6
B. $\frac{1}{4}$
C. $\frac{1}{3}$
D. $\frac{1}{2}$

17) If the perimeter of the following figure be 33, what is the value of x

A. 4
B. 6
C. 8
D. 11

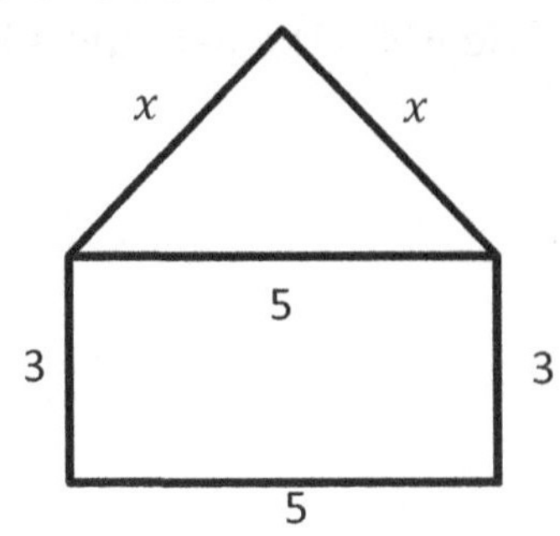

18) Right triangle ABC has two legs of lengths $3\ cm$ (AB) and $4\ cm$ (AC). What is the length of the third side (BC)?

A. $10\ cm$
B. $8\ cm$
C. $6\ cm$
D. $5\ cm$

19) What is the equivalent temperature of $140°F$ in Celsius?

$$C = \frac{5}{9}(F - 32)$$

A. 32
B. 60
C. 68
D. 72

20) Simplify $7x^2y^3(2x^2y)^3 =$

A. $12x^4y^6$
B. $12x^8y^6$
C. $56x^4y^6$
D. $56x^8y^6$

STOP: This is the End of Section 2 of test 2.

Next Generation Accuplacer Mathematics Practice Test 2

2019 - 2020

Section 3: Advanced Algebra and Functions

(Calculator)

20 questions

Total time for this section: No time limit.

You may use a calculator on this Section.

(On a real Accuplacer test, there is an onscreen calculator to use on some questions.)

1) If $\frac{x-3}{5} = N$ and $N = 7$, what is the value of x?
   A. 25
   B. 28
   C. 30
   D. 38

2) Which of the following is equal to $b^{\frac{4}{5}}$?
   A. $\sqrt{b^{\frac{4}{5}}}$
   B. $b^{\frac{4}{5}}$
   C. $\sqrt[5]{b^4}$
   D. $\sqrt[4]{b^5}$

3) Solve the following equation for y?

$$\frac{x}{2+3} = \frac{y}{10-7}$$

   A. $\frac{3}{5}x$
   B. $\frac{5}{3}x$
   C. $2x$
   D. $3x$

4) For $i = \sqrt{-1}$, which of the following is equivalent of $\frac{2+3i}{5-2i}$?
   A. $\frac{3+2i}{5}$
   B. 5+3i
   C. $\frac{4+19i}{29}$
   D. $\frac{4+19i}{20}$

5) If function is defined as $f(x) = bx^2 + 15$, and b is a constant and $f(2) = 35$. What is the value of $f(4)$?
   A. 25
   B. 45
   C. 95
   D. 105

6) Find the solution (x, y) to the following system of equations?

$$-3x - y = 6$$
$$6x + 4y = 10$$

A. $(14, 5)$
B. $(6, 8)$
C. $(11, 17)$
D. $(-\frac{17}{3}, 11)$

7) Calculate $f(3)$ for the function $f(x) = 3x^2 - 4$.
A. 23
B. 30
C. 48
D. 50

8) What is the sum of all values of n that satisfies $2n^2 + 16n + 24 = 0$?
A. 8
B. 4
C. -4
D. -8

9) For $i = \sqrt{-1}$, what is the value of $\frac{3+2i}{5+i}$?
A. i
B. $\frac{32i}{5}$
C. $\frac{17-i}{5}$
D. $\frac{17+7i}{26}$

$$y = x^2 - 8x + 12$$

10) The equation above represents a parabola in the xy-plane. Which of the following equivalent forms of the equation displays the x-intercepts of the parabola as constants or coefficients?
A. $y = x + 3$
B. $y = x(x - 8)$
C. $y = (x + 6)(x + 2)$
D. $y = (x - 6)(x - 2)$

11) The function $g(x)$ is defined by a polynomial. Some values of x and $g(x)$ are shown in the table below. Which of the following must be a factor of $g(x)$?

A. x
B. $x + 1$
C. $x - 1$
D. $x + 3$

x	$g(x)$
0	3
1	0
2	-3
3	-8
4	-12

12) What is the value of $\frac{6b}{c}$ when $\frac{c}{b} = 2$

A. 6
B. 4
C. 3
D. 1

13) Which of the following is equivalent to $\frac{x+(5x)^2+(3x)^3}{x}$?

A. $16x^2 + 25x + 1$
B. $27x^2 + 25x + 1$
C. $16x^2 + 25x$
D. $27x^3 + 16x^2 + 1$

14) If $\frac{a-b}{b} = \frac{10}{11}$, then which of the following must be true?

A. $\frac{a}{b} = \frac{11}{10}$
B. $\frac{a}{b} = \frac{21}{11}$
C. $\frac{a}{b} = \frac{11}{21}$
D. $\frac{a}{b} = \frac{21}{10}$

15) Which of the following lines is parallel to: $6y - 2x = 30$

A. $y = \frac{1}{3}x + 2$
B. $y = 3x + 5$
C. $y = x - 2$
D. $y = 2x - 1$

16) What is the average of $4x + 2, -6x - 5$ and $8x + 2$?

A. $3x + 2$
B. $3x - 2$
C. $2x + 1$
D. $2x - \frac{1}{3}$

17) A rectangle was altered by increasing its length by 20 percent and decreasing its width by s percent. If these alterations decreased the area of the rectangle by 4 percent, what is the value of s.

A. 40
B. 31
C. 20
D. 10

18) Tickets for a talent show cost $\$3$ for children and $\$4$ for adults. If John spends at least $\$10$ but no more than $\$15$ on x children tickets and 2 adult ticket, what are two possible values of x?

A. $0, 1$
B. $1, 2$
C. $1, 4$
D. $1, 5$

19) If the interior angles of a quadrilateral are in the ratio $1:4:6:7$, what is the measure of the smallest angle?

A. 18°
B. 20°
C. 72°
D. 120°

20) A plant grows at a linear rate. After five weeks, the plant is $45\ cm$ tall. Which of the following functions represents the relationship between the height (y) of the plant and number of weeks of growth (x)?

A. $y(x) = 45x + 9$
B. $y(x) = 9x + 45$
C. $y(x) = 45x$
D. $y(x) = 9x$

STOP: This is the End of Test 2.

ACCUPLACER Mathematics Practice Test Answers and Explanations

Now, it's time to review your results to see where you went wrong and what areas you need to improve!

Accuplacer Math Test 1											
Arithmetic				Quantitative Reasoning, Algebra, And Statistics				Advanced Algebra and Functions			
1	**D**	16	**C**	1	**C**	16	**B**	1	**C**	16	**B**
2	**A**	17	**C**	2	**A**	17	**B**	2	**A**	17	**A**
3	**D**	18	**B**	3	**C**	18	**B**	3	**C**	18	**C**
4	**B**	19	**C**	4	**C**	19	**D**	4	**C**	19	**C**
5	**C**	20	**D**	5	**C**	20	**A**	5	**C**	20	**B**
6	**C**			6	**A**			6	**A**		
7	**C**			7	**D**			7	**B**		
8	**A**			8	**D**			8	**D**		
9	**D**			9	**C**			9	**B**		
10	**B**			10	**D**			10	**B**		
11	**C**			11	**C**			11	**C**		
12	**D**			12	**B**			12	**D**		
13	**C**			13	**C**			13	**C**		
14	**B**			14	**B**			14	**C**		
15	**C**			15	**C**			15	**C**		

Accuplacer Math Test 2

Arithmetic

1	B	16	D
2	C	17	D
3	D	18	C
4	C	19	D
5	A	20	C
6	A		
7	D		
8	A		
9	C		
10	A		
11	A		
12	D		
13	D		
14	B		
15	A		

Quantitative Reasoning, Algebra, And Statistics

1	A	16	D
2	A	17	D
3	A	18	D
4	C	19	B
5	D	20	D
6	D		
7	B		
8	B		
9	D		
10	B		
11	A		
12	B		
13	D		
14	B		
15	C		

Advanced Algebra and Functions

1	D	16	D
2	C	17	C
3	A	18	B
4	C	19	B
5	C	20	D
6	D		
7	A		
8	D		
9	D		
10	D		
11	C		
12	C		
13	B		
14	B		
15	A		

ACCUPLACER Mathematics Practice Test 1

Arithmetic

1) Choice D is correct

To compare fractions, find a common denominator. When two fractions have common denominators, the fraction with the larger numerator is the larger number. Choice A is incorrect because $\frac{3}{4}$ is not less than $\frac{17}{24}$. Write both fractions with common denominator and compare the numerators. $\frac{3}{4}=\frac{18}{24}$. The fraction $\frac{18}{24}$ is greater than $\frac{17}{24}$.

Choice B and C are not correct. Shown written with a common denominator, the comparisons $\frac{2}{3}<\frac{7}{9}$ and $\frac{3}{8}<\frac{9}{25}$ are not correct. Shown written with a common denominator, the comparison $\frac{11}{21}<\frac{4}{7}$ is correct because $\frac{4}{7}$ or $\frac{12}{21}$ is greater than $\frac{11}{21}$.

2) Choice A is correct

First multiply the tenths place of 7.8 by 4.56. The result is 3.648. Next, multiply 7 by 4.56 which results in 31.92. The sum of these two numbers is: $3.648 + 31.92 = 35.568$

3) Choice D is correct

Dividing 72 by 12%, which is equivalent to 0.12, gives 600. Therefore, 12% of 600 is 72.

4) Choice B is correct.

$\frac{13}{8} = 1.6$, the only choice that is greater than 1.6 is $\frac{5}{2}$. $\frac{5}{2} = 2.5\,, 2.5 > 1.6$

5) Choice C is correct

$\$8 \times 10 = \$80$, Petrol use: $10 \times 2 = 20$ liters

Petrol cost: $20 \times \$1 = \$20$. Money earned: $\$80 - \$20 = \$60$

6) Choice C is correct.

The closest to 6.03 is 6 in the options provided.

7) Choice C is correct

Let x be the original price. If the price of the sofa is decreased by 20% to \$476, then: $80\ \%\ of\ x = 476 \Rightarrow 0.80x = 476 \Rightarrow x = 476 \div 0.80 = 595$

8) Choice A is correct

The percent of girls playing tennis is: $45\% \times 25\% = 0.45 \times 0.25 = 0.11 = 11\%$

9) Choice D is correct

Write the equation and solve for B: $0.60\ A = 0.30\ B$, divide both sides by 0.30, then: $\frac{0.60}{0.30}A = B$, therefore: $B = 2A$, and B is 2 times of A or it's 200% of A.

10) Choice B is correct

Use this formula: Percent of Change $\frac{New\ Value - Old\ Value}{Old\ Value} \times 100\%$

$\frac{16000-20000}{20000} \times 100\% = -20\%$ and $\frac{12800-16000}{16000} \times 100\% = -20\%$

11) Choice C is correct

Use simple interest formula: $I = prt$ $(I = interest,\ p = principal, r = rate, t = time)$

$I = (9{,}000)(0.035)(5) = 1{,}575$

12) Choice D is correct.

First simplify the multiplication: $\frac{5}{4} \times \frac{6}{2} = \frac{30}{8} = \frac{15}{4}$, Choice D is equal to $\frac{15}{4}$. $\frac{5\times3}{4} = \frac{15}{4}$

13) Choice C is correct

Use percent formula: $part = \frac{percent}{100} \times whole$

$35 = \frac{\text{percent}}{100} \times 20 \Rightarrow 35 = \frac{\text{percent} \times 20}{100} \Rightarrow 35 = \frac{\text{percent} \times 2}{10}$, multiply both sides by 10.

$350 = percent \times 2$, divide both sides by 2. $175 = percent$

14) Choice B is correct

To add decimal numbers, line them up and add from right.

$3.85 + 0.045 + 0.1365 = 4.0315$

15) Choice C is correct

Use distance formula: $Distance = Rate \times time \Rightarrow 420 = 65 \times T$, divide both sides by 65.
$420 \div 65 = T \Rightarrow T = 6.4\ hours.$

Change hours to minutes for the decimal part. $0.4\ hours = 0.4 \times 60 = 24\ minutes.$

16) Choice C is correct

Let's compare each fraction: $\frac{2}{7} < \frac{3}{8} < \frac{5}{11} < \frac{5}{4}$ Only choice C provides the right order.

17) Choice C is correct.

$\frac{4}{5} - \frac{2}{5} = \frac{2}{5} = 0.4$

18) Choice B is correct

$\frac{(8+6)^2}{2} + 6 = \frac{(14)^2}{2} + 6 = \frac{196}{2} + 6 = 98 + 6 = 104$

19) Choice C is correct

The weight of 14.2 meters of this rope is: $14.2 \times 600\ g = 8{,}520\ g$

$1\ kg = 1{,}000\ g$, therefore, $8{,}520\ g \div 1000 = 8.52\ kg$

20) Choice D is correct.

78 divided by 5, the remainder is 3. 45 divided by 7, the remainder is also 3.

Accuplacer Mathematics Practice Test 1

Quantitative Reasoning, Algebra, And Statistics

1) Choice C is correct

Let x be the number. Write the equation and solve for x. $(28 - x) \div x = 3$

Multiply both sides by x. $(28 - x) = 3x$, then add x both sides. $28 = 4x$, now divide both sides by 4. $x = 7$

2) Choice A is correct

The sum of supplement angles is 180. Let x be that angle. Therefore, $x + 5x = 180$

$6x = 180$, divide both sides by 6: $x = 30$

3) Choice C is correct

The average speed of john is: $150 \div 5 = 30\ km$, The average speed of Alice is: $180 \div 4 = 45\ km$, Write the ratio and simplify. $30 : 45 = 2 : 3$

4) Choice C is correct

$\$7 \times 10 = \$70$, Petrol use: $10 \times 2 = 20$ liters

Petrol cost: $20 \times \$1 = \$20$, Money earned: $\$70 - \$20 = \$50$

5) Choice C is correct

Use Pythagorean Theorem: $a^2 + b^2 = c^2$

$6^2 + 8^2 = c^2 \Rightarrow 36 + 64 = c^2 \Rightarrow 100 = c^2 \Rightarrow \text{c} = 10$

6) Choice A is correct

Area of the circle is less than 14π. Use the formula of areas of circles.

$Area = \pi r^2 \Rightarrow 49\,\pi > \pi r^2 \Rightarrow 49 > r^2 \Rightarrow r < 7$

Radius of the circle is less than 7. Let's put 7 for the radius. Now, use the circumference formula: $Circumference = 2\pi r = 2\pi\ (7) = 14\ \pi$

Since the radius of the circle is less than 7. Then, the circumference of the circle must be less than 14π. Only choice A is less than 14π.

7) Choice D is correct

If the length of the box is 36, then the width of the box is one third of it, 12, and the height of the box is 4 (one third of the width). The volume of the box is:

$V = lwh = (36)(12)(4) = 1{,}728$

8) Choice D is correct

To find the number of possible outfit combinations, multiply number of options for each factor: $6 \times 4 \times 5 = 120$

9) Choice C is correct

The area of trapezoid is: $\left(\frac{8+12}{2}\right) \times x = 100 \to 10x = 100 \to x = 10$

$y = \sqrt{3^2 + 4^2} = 5$, Perimeter is: $12 + 10 + 8 + 5 = 35$

10) Choice D is correct

The equation of a line is in the form of $y = mx + b$, where m is the slope of the line and b is the $y - intercept$ of the line. Two points (4,3) and (3,2) are on line A. Therefore, the slope of the line A is: $slope\ of\ line\ A = \frac{y_2 - y_1}{x_2 - x_1} = \frac{2-3}{3-4} = \frac{-1}{-1} = 1$

The slope of line A is 1. Thus, the formula of the line A is: $y = mx + b = x + b$, choose a point and plug in the values of x and y in the equation to solve for b. Let's choose point $(4, 3)$. Then: $y = x + b \to 3 = 4 + b \to b = 3 - 4 = -1$. The equation of line A is:

$$y = x - 1$$

Now, let's review the choices provided:

A. $(-1, 2)$ $\quad y = x - 1 \to 2 = -1 - 1 = -2$ This is not true.

B. $(5, 7)$ $\quad y = x - 1 \to 7 = 5 - 1 = 4$ This is not true.

C. $(3, 4)$ $\quad y = x - 1 \to 4 = 3 - 1 = 2$ This is not true.

D. $(-1, -2)$ $\quad y = x - 1 \to -2 = -1 - 1 = -2$ This is true!

11) Choice C is correct

Let x be the number. Write the equation and solve for x.

$\frac{2}{3} \times 15 = \frac{2}{5} \cdot x \Rightarrow \frac{2\times15}{3} = \frac{2x}{5}$, use cross multiplication to solve for x.

$5 \times 30 = 2x \times 3 \Rightarrow 150 = 6x \Rightarrow x = 25$

12) Choice B is correct

Use the information provided in the question to draw the shape.

120 miles

50 miles

Use Pythagorean Theorem: $a^2 + b^2 = c^2$

$50^2 + 120^2 = c^2 \Rightarrow 2500 + 14400 = c^2 \Rightarrow 16900 = c^2 \Rightarrow c = 130$

13) Choice C is correct

The ratio of boy to girls is 4: 7. Therefore, there are 4 boys out of 11 students. To find the answer, first divide the total number of students by 11, then multiply the result by 4.

$55 \div 11 = 4 \Rightarrow 5 \times 4 = 20$ There are 20 boys and 35 $(55 - 20)$ girls. So, 15 more boys should be enrolled to make the ratio $1:1$

14) Choice B is correct

If the score of Mia was 40, therefore the score of Ava is 20. Since, the score of Emma was half as that of Ava, therefore, the score of Emma is 10.

15) Choice C is correct

The rate of construction company $= \frac{40\ cm}{1\ min} = 40\frac{cm}{min}$

Height of the wall after 50 $min = \frac{40\ cm}{1\ min} \times 50\ min = 2{,}000cm$

Let x be the height of wall, then $\frac{2}{3}x = 2{,}000\ cm \rightarrow x = \frac{3 \times 2{,}000}{2} \rightarrow x = 3{,}000\ cm = 30m$

16) Choice B is correct

Let x be the smallest number. Then, these are the numbers: $x, x+1, x+2, x+3, x+4$

$average = \frac{sum\ of\ terms}{number\ of\ terms} \Rightarrow 40 = \frac{x+(x+1)+(x+2)+(x+3)+(x+4)}{5} \Rightarrow 40 = \frac{5x+10}{5} \Rightarrow 200 = 5x + 10$
$\Rightarrow 190 = 5x \Rightarrow x = 38$

17) Choice B is correct

$average = \frac{sum\ of\ terms}{number\ of\ terms}$, The sum of the weight of all girls is: $18 \times 55 = 990\ kg$

The sum of the weight of all boys is: $32 \times 62 = 1{,}984\ kg$, The sum of the weight of all students is: $990 + 1{,}984 = 2{,}974\ kg$, $average = \frac{2{,}974}{50} = 59.48$

18) Choice B is correct

Write the numbers in order: $3, 5, 8, 9, 12, 15, 19$. Since we have 7 numbers (7 is odd), then the median is the number in the middle, which is 9.

19) Choice D is correct

Formula for the Surface area of a cylinder is: $SA = 2\pi r^2 + 2\pi rh$

$\rightarrow 150\pi = 2\pi r^2 + 2\pi r(10) \rightarrow r^2 + 10r - 75 = 0$

$(r + 15)(r - 5) = 0 \rightarrow r = 5 \quad or \quad r = -15\ (unacceptable)$

20) Choice A is correct

Let x be the number of years. Therefore, \$2,000 per year equals $2{,}000x$. Starting from \$25,000 annual salary means you should add that amount to $2{,}000x$.

Income more than that is: $I > 2{,}000x + 25{,}000$

Accuplacer Mathematics Practice Test 1

Advanced Algebra and Functions

1) Choice C is correct

Method 1: Plugin the values of x and y provided in the options into both equations.

A. $(4,3)$ $x+y=0 \rightarrow 4+3 \neq 0$
B. $(5,4)$ $x+y=0 \rightarrow 5+4 \neq 0$
C. $(4,-4)$ $x+y=0 \rightarrow 4+(-4)=0$
D. $(4,-6)$ $x+y=0 \rightarrow 4+(-6) \neq 0$

Only option C is correct.

Method 2: Multiplying each side of $x+y=0$ by 2 gives $2x+2y=0$. Then, adding the corresponding side of $2x+2y=0$ and $4x-2y=24$ gives $6x=24$. Dividing each side of $6x=24$ by 6 gives $x=4$. Finally, substituting 4 for x in $x+y=0$, or $y=-4$. Therefore, the solution to the given system of equations is $(4,-4)$.

2) Choice A is correct

If $f(x)=3x+4(x+1)+2$, then find $f(3x)$ by substituting $3x$ for every x in the function. This gives: $f(3x)=3\,(3x)+\ 4(3x+1)+2$

It simplifies to: $f(3x)=3\,(3x)+4(3x+1)+2=9x+12x+4+2=21x+6$

3) Choice C is correct

First, find the equation of the line. All lines through the origin are of the form $y=mx$, so the equation is $y=\frac{1}{3}x$. Of the given choices, only choice C (9,3), satisfies this equation:

$y=\frac{1}{3}x \rightarrow 3=\frac{1}{3}(9)=3$

4) Choice C is correct

$(3n^2+4n+6)-(2n^2-5)$. Add like terms together: $3n^2-2n^2=n^2$

$4n$ doesn't have like terms. $6-(-5)=11$

Combine these terms into one expression to find the answer: $n^2+4n+11$

5) Choice C is correct

You can find the possible values of a and b in $(ax+4)(bx+3)$ by using the given equation $a+b=7$ and finding another equation that relates the variables a and b. Since $(ax+4)(bx+3)=10x^2+cx+12$, expand the left side of the equation to obtain

$abx^2+4bx+3ax+12=10x^2+cx+12$

Since ab is the coefficient of x^2 on the left side of the equation and 10 is the coefficient of x^2 on the right side of the equation, it must be true that $ab=10$

The coefficient of x on the left side is $4b + 3a$ and the coefficient of x in the right side is c.
Then: $4b + 3a = c, \quad a + b = 7$, then: $a = 7 - b$

Now, plug in the value of a in the equation $ab = 10$. Then:

$ab = 10 \rightarrow (7 - b)b = 10 \rightarrow 7b - b^2 = 10$

Add $-7b + b^2$ both sides. Then: $b^2 - 7b + 10 = 0$

Solve for b using the factoring method. $b^2 - 7b + 10 = 0 \rightarrow (b - 5)(b - 2) = 0$

Thus, either $b = 2$ and $a = 5$, or $b = 5$ and $a = 2$. If $b = 2$ and $a = 5$, then

$4b + 3a = c \rightarrow 4(2) + 3(5) = c \rightarrow c = 23$. If $5 = 2$ and $a = 2$, then, $4b + 3a = c \rightarrow 4(5) + 3(2) = c \rightarrow c = 26$. Therefore, the two possible values for c are 23 and 26.

6) Choice A is correct

To rewrite $\frac{1}{\frac{1}{x-6}+\frac{1}{x+4}}$, first simplify $\frac{1}{x-6} + \frac{1}{x+4}$.

$\frac{1}{x-6} + \frac{1}{x+4} = \frac{1(x+4)}{(x-6)(x+4)} + \frac{1(x-5)}{(x+4)(x-6)} = \frac{(x+4)+(x-6)}{(x+4)(x-6)}$

Then: $\frac{1}{\frac{1}{x-6}+\frac{1}{x+4}} = \frac{1}{\frac{(x+4)+(x-6)}{(x+4)(x-6)}} = \frac{(x-6)(x+4)}{(x-6)+(x+4)}$. (Remember, $\frac{1}{\frac{1}{x}} = x$)

This result is equivalent to the expression in choice A.

7) Choice B is correct

Since $(0, 0)$ is a solution to the system of inequalities, substituting 0 for x and 0 for y in the given system must result in two true inequalities. After this substitution, $y < a - x$ becomes $0 < a$, and $y > x + b$ becomes $0 > b$. Hence, a is positive and b is negative. Therefore, $a > b$.

8) Choice D is correct

First find the slope of the line using the slope formula. $m = \frac{y_2 - y_1}{x_2 - x_1}$

Substituting in the known information. $(x_1, y_1) = (2, 4), \quad (x_2, y_2) = (4,5)$

$m = \frac{5-4}{4-2} = \frac{1}{2}$

Now the slope to find the equation of the line passing through these points. $y = mx + b$

Choose one of the points and plug in the values of x and y in the equation to solve for b.

Let's choose point $(4, 5)$. Then: $y = mx + b \rightarrow 5 = \frac{1}{2}(4) + b \rightarrow 5 = 2 + b \rightarrow b = 5 - 2 = 3$

The equation of the line is: $y = \frac{1}{2}x + 3$

Now, plug in the points provided in the choices into the equation of the line.

A. $(9, 9)$ $\quad y = \frac{1}{2}x + 3 \rightarrow 9 = \frac{1}{2}(9) + 3 \rightarrow 9 = 7.5$ This is NOT true.

B. $(9, 6)$ $\quad y = \frac{1}{2}x + 3 \rightarrow 6 = \frac{1}{2}(9) + 3 \rightarrow 6 = 7.5$ This is NOT true.

C. $(6,9)$ $\quad y = \frac{1}{2}x + 3 \rightarrow 9 = \frac{1}{2}(6) + 3 \rightarrow 9 = 6$ This is NOT true.

D. $(6,6)$ $\quad y = \frac{1}{2}x + 3 \rightarrow 6 = \frac{1}{2}(6) + 3 \rightarrow 6 = 6$ This is true!

Therefore, the only point from the choices that lies on the line is $(6,6)$.

9) Choice B is correct

The input value is 4. Then: $x = 4$

$f(x) = x^2 - 3x \rightarrow f(4) = 4^2 - 3(4) = 16 - 12 = 4$

10) Choice B is correct

To solve this problem, first recall the equation of a line: $y = mx + b$

Where $m = slope.$ $\quad y = y - intercept$

Remember that slope is the rate of change that occurs in a function and that the y −intercept is the y value corresponding to $x = 0$. Since the height of John's plant is 6 inches tall when he gets it. Time (or x) is zero. The plant grows 4 inches per year. Therefore, the rate of change of the plant's height is 4. The y −intercept represents the starting height of the plant which is 6 inches.

11) Choice C is correct

Multiplying each side of $\frac{3}{x} = \frac{12}{x-9}$ by $x(x-9)$ gives $3(x-9) = 12(x)$, distributing the 3 over the values within the parentheses yields $x - 9 = 4x$ or $x = -3$.

Therefore, the value of $\frac{x}{6} = \frac{-3}{6} = -\frac{1}{2}$.

12) Choice D is correct

The equation of a circle can be written as $(x-h)^2 + (y-k)^2 = r^2$ where (h,k) are the coordinates of the center of the circle and r is the radius of the circle. Since the coordinates of the center of the circle are $(0,4)$, the equation is $x^2 + (y-4)^2 = r^2$, where r is the radius. The radius of the circle is the distance from the center $(0,4)$, to the given endpoint of a radius, $\left(\frac{5}{3}, 6\right)$. By the distance formula, $r^2 = \left(\frac{5}{3} - 0\right)^2 + (6-4)^2 = \frac{61}{9}$

Therefore, an equation of the given circle is $x^2 + (y-4)^2 = \frac{61}{9}$

13) Choice C is correct

To solve for $\cos A$ first identify what is known. The question states that ΔABC is a right triangle whose $n\angle B = 90^\circ$ and $\sin C = \frac{2}{3}$. It is important to recall that any triangle has a sum of interior angles that equals 180 degrees. Therefore, to calculate $\cos A$ use the complimentary angles identify of trigonometric function. $\cos A = \cos(90 - C)$, Then: $\cos A = \sin C$

For complementary angles, sin of one angle is equal to cos of the other angle. $\cos A = \frac{2}{3}$

14) Choice C is correct

In order to figure out what the equation of the graph is, fist find the vertex. From the graph we can determine that the vertex is at (1,2). We can use vertex form to solve for the equation of this graph. Recall vertex form, $y = a(x-h)^2 + k$, where h is the x coordinate of the vertex, and k is the y coordinate of the vertex. Plugging in our values, you get $= a(x-1)^2 + 2$, To solve for a, we need to pick a point on the graph and plug it into the equation. Let's pick $(-1, 10)$, $10 = a(-1-1)^2 + 2$

$10 = a(-2)^2 + 2,$ $\quad 10 = 4a + 2,$ $\quad 8 = 4a,$ $\quad a = 2$

Now the equation is : $y = 2(x-1)^2 + 2$

Let's expand this, $y = 2(x^2 - 2x + 1) + 2$, $y = 2x^2 - 4x + 2 + 2$

$y = 2x^2 - 4x + 4.$ The equation in Choice C is the same.

15) Choice C is correct

The line passes through the origin, $(6, m)$ and $(m, 12)$. Any two of these points can be used to find the slope of the line. Since the line passes through $(0,0)$ and $(6, m)$, the slope of the line is equal to $\frac{m-0}{6-0} = \frac{m}{6}$. Similarly, since the line passes through $(0,0)$ and $(m, 12)$, the slope of the line is equal to $\frac{12-0}{m-0} = \frac{12}{m}$. Since each expression gives the slope of the same line, it must be true that $\frac{m}{6} = \frac{12}{m}$,Using cross multiplication gives

$\frac{m}{6} = \frac{12}{m} \rightarrow m^2 = 72 \rightarrow m = \pm\sqrt{72} = \pm\sqrt{36 \times 2} = \pm\sqrt{36} \times \sqrt{2} = \pm 6\sqrt{2}$

16) Choice B is correct

It is given that $g(6) = 4$. Therefore, to find the value of $f(g(6))$, then $f(g(6)) = f(4) = 7$

17) Choice A is correct

Area of the triangle is: $\frac{1}{2}\ AD \times BC$ and AD is perpendicular to BC. Triangle ADC is a $30° - 60° - 90°$ right triangle. The relationship among all sides of right triangle $30° - 60° - 90°$ is provided in the following triangle: In this triangle, the opposite side of $30°$ angle is half of the hypotenuse. And the opposite side of $60°$ is opposite of $30° \times \sqrt{3}$

$CD = 4$, then $AD = 4 \times \sqrt{3}$

Area of the triangle ABC is: $\frac{1}{2}\ AD \times BC = \frac{1}{2}\ 4\sqrt{3} \times 8 = 16\sqrt{3}$

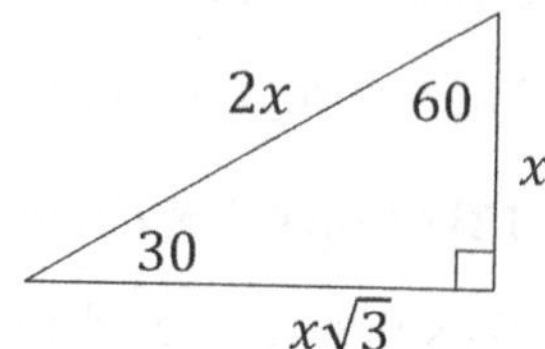

18) Choice C is correct

It is given that $g(7) = 8$. Therefore, to find the value of $f(g(7))$, substitute 8 for $g(7)$. $f(g(7)) = f(8) = 35$.

19) Choice C is correct

The equation of a circle in standard form is: $(x-h)^2 + (y-k)^2 = r^2$, where r is the radius of the circle. In this circle the radius is 4. $r^2 = 16 \rightarrow r = 4$, $(x+2)^2 + (y-4)^2 = 16$

Area of a circle: $A = \pi r^2 = \pi(4)^2 = 16\pi$

20) Choice B is correct

By definition, the sine of any acute angle is equal to the cosine of its complement. Since, angle A and B are complementary angles, therefore: $sin\ A = cos\ B$, it's time to review your results to see where you went wrong and what areas you need to improve!

ACCUPLACER Mathematical Reasoning Practice Test 2

Arithmetic

1) Choice B is correct.

$5 \text{ percent of } 560 = \frac{5}{100} \times 560 = \frac{1}{20} \times 560 = \frac{560}{20} = 28$

2) Choice C is correct

The population is increased by 10% and 20%. 10% increase changes the population to 110% of original population. For the second increase, multiply the result by 120%.

$(1.10) \times (1.20) = 1.32 = 132\%$. 32 percent of the population is increased after two years.

3) Choice D is correct.

$75 off is the same as 20 percent off. Thus, 20 percent of a number is 75.

Then: $20\%\ of\ x = 75 \rightarrow 0.2x = 75 \rightarrow x = \frac{75}{0.2} = 375$

4) Choice C is correct

Three times of 25,000 is 75,000. One sixth of them cancelled their tickets. One sixth of 75,000 equals 12,500 $(\frac{1}{6} \times 75{,}000 = 12{,}500)$. $62{,}500(75{,}000 - 12{,}500 = 62{,}500)$ fans are attending this week.

5) Choice A is correct

$$\frac{1\frac{5}{4}+\frac{1}{3}}{2\frac{1}{2}-\frac{15}{8}} = \frac{\frac{9}{4}+\frac{1}{3}}{\frac{5}{2}-\frac{15}{8}} = \frac{\frac{27+4}{12}}{\frac{20-15}{8}} = \frac{\frac{31}{12}}{\frac{5}{8}} = \frac{31 \times 8}{12 \times 5} = \frac{31 \times 2}{3 \times 5} = \frac{62}{15} \cong 4.133$$

6) Choice A is correct

To find the sum of two whole numbers, line the numbers up and add digits from right.

$$\begin{array}{r} 12{,}181 \\ +\ 8{,}951 \\ \hline 21{,}132 \end{array}$$

7) Choice D is correct

20% of $150 is $0.2 \times 150 = 30$

8) Choice A is correct

First, find the number. Let x be the number. 150% of a number is 75, then: $1.5 \times x = 75 \Rightarrow x = 75 \div 1.5 = 50$, 80% of 50 is: $0.8 \times 50 = 40$

9) Choice C is correct

The result when 754 is divided by 7 is 107 with a remainder of 5. Multiplying $7 \times 107 = 749$ and $754 - 749 = 5$, which is the remainder.

10) Choice A is correct.

$\frac{3}{2} = 1.5$. The only choice less than 1.5 is 1.3. $\frac{3}{2} = 1.5 > 1.3$

11) Choice A is correct

2,500 out of 65,000 equals to $\frac{2500}{65000} = \frac{25}{650} = \frac{1}{26}$

12) Choice D is correct

Dividing 75 by 15%, which is equivalent to 0.15, gives 500.

13) Choice D is correct

The failing rate is 11 out of 44 = $\frac{11}{44}$. Change the fraction to percent: $\frac{11}{44} \times 100\% = 25\%$

25 percent of students failed. Therefore, 75 percent of students passed the exam.

14) Choice B is correct

Use simple interest formula: $I = prt$ (I = interest, p = principal, r = rate, t = time)

$I = (13{,}000)(0.035)(2) = 910$

15) Choice A is correct.

The second digit to the right of the decimal point is in the hundredths place and the third number to the right of the decimal point is in the thousandths place. Since the number in the thousandths place of 0.6749, which is 4, is less than 5, the number 0.6749 should be rounded down to 0.67.

16) Choice D is correct

$\frac{3}{4} \times 28 = \frac{84}{4} = 21$

17) Choice D is correct

The amount of money that jack earns for one hour: $\frac{\$616}{44} = \$14$

Number of additional hours that he work to make enough money is: $\frac{\$826-\$616}{1.5\times\$14} = 10$

Number of total hours is: $44 + 10 = 54$

18) Choice C is correct

The second digit to the right of the decimal point is in the hundredths place and the third number to the right of the decimal point is in the thousandths place. Since the number in the thousandths place of 420.756 , which is 6, is greater than 5, the number 420.756 should be rounded up to 420.76.

19) Choice D is correct

To best compare the numbers, they should be put in the same format. The percent 0.625% can be converted to a decimal by dividing 0.625 by 100, which gives 0.00625. $\frac{5}{8}$ can be converted to a decimal by dividing 5 by 8, which gives 0.625. Now, all three numbers are in decimal format. $0.625 > 0.0625 > 0.00625$ or $\frac{5}{8} > 0.0625 > 0.625\%$, which is choice D.

20) Choice C is correct

The fraction $\frac{3}{5}$ can be written as $\frac{3\times20}{5\times20} = \frac{60}{100}$, which can be interpreted as forty hundredths, or 0.60.

Accuplacer Mathematics Practice Test 2

Quantitative Reasoning, Algebra, And Statistics

1) Choice A is correct

In the stadium the ratio of home fans to visiting fans in a crowd is $4:8$. Therefore, total number of fans must be divisible by $12: 5 + 7 = 12$.

Let's review the choices:

A. $12{,}324$: $12{,}324 \div 12 = 1{,}027$

B. $42{,}326$ $42{,}326 \div 12 = 3{,}527.166$

C. $44{,}566$ $44{,}566 \div 12 = 3{,}713.833$

D. $66{,}812$ $66{,}812 \div 12 = 5{,}567.666$

Only choice A when divided by 12 results a whole number.

2) Choice A is correct

$x + 2y = 4$. Plug in the values of x and y from choices provided. Then:

A. $(-2,3)$ $x + 2y = 4 \rightarrow -2 + 2(3) = 4 \rightarrow -2 + 6 = 4$ This is true!

B. $(1,2)$ $x + 2y = 4 \rightarrow 1 + 2(2) = 4 \rightarrow 1 + 4 = 4$ This is NOT true!

C. $(-1,3)$ $x + 2y = 4 \rightarrow -1 + 2(3) = 4 \rightarrow -1 + 6 = 4$ This is NOT true!

D. $(-3,4)$ $x + 2y = 4 \rightarrow -3 + 2(4) = 4 \rightarrow -3 + 8 = 4$ This is NOT true!

3) Choice A is correct

The width of the rectangle is twice its length. Let x be the length. Then, $width = 2x$

Perimeter of the rectangle is $2(width + length) = 2(2x + x) = 72 \Rightarrow 6x = 72 \Rightarrow x = 12$

Length of the rectangle is 12 meters.

4) Choice C is correct

$average\ (mean) = \frac{sum\ of\ terms}{number\ of\ terms} \Rightarrow 86 = \frac{\text{sum of terms}}{50} \Rightarrow sum = 86 \times 50 = 4{,}300$

The difference of 91 and 66 is 25. Therefore, 25 should be subtracted from the sum.

$4{,}300 - 25 = 4{,}275,\ mean = \frac{sum\ of\ terms}{number\ of\ terms} \Rightarrow \text{mean} = \frac{4{,}275}{50} = 85.5$

5) Choice D is correct

We have two equations and three unknown variables, therefore x cannot be obtained.

6) Choice D is correct

Use formula of rectangle prism volume. $V = (length)(width)(height) \Rightarrow 2{,}500 = (25)(10)(height) \Rightarrow height = 2{,}500 \div 250 = 10$

7) Choice B is correct

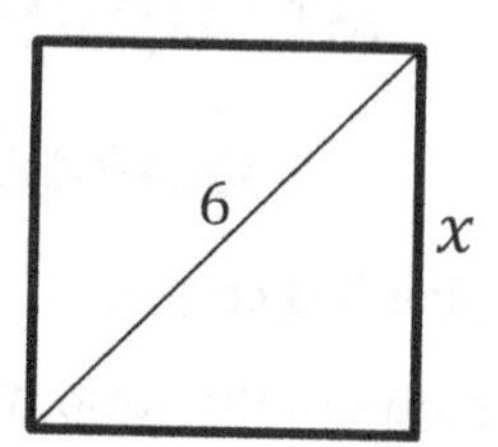

The diagonal of the square is 6. Let x be the side.

Use Pythagorean Theorem: $a^2 + b^2 = c^2$

$x^2 + x^2 = 6^2 \Rightarrow 2x^2 = 6^2 \Rightarrow 2x^2 = 36 \Rightarrow x^2 = 18 \Rightarrow x = \sqrt{18}$

The area of the square is: $\sqrt{18} \times \sqrt{18} = 18$

8) Choice B is correct

$Probability = \frac{number\ of\ desired\ outcomes}{\text{number of total outcomes}} = \frac{10}{15+10+10+25} = \frac{10}{60} = \frac{1}{6}$

9) Choice D is correct

$average = \frac{sum\ of\ terms}{number\ of\ terms} \Rightarrow$ (average of 6 numbers) $14 = \frac{\text{sum of numbers}}{6} \Rightarrow$sum of 6 numbers is: $14 \times 6 = 84$

(average of 4 numbers) $10 = \frac{\text{sum of numbers}}{4} \Rightarrow$sum of 4 numbers is $10 \times 4 = 40$

$sum\ of\ 6\ numbers - sum\ of\ 4\ numbers = sum\ of\ 2\ numbers \rightarrow 84 - 40 = 44$

average of 2 numbers $= \frac{44}{2} = 22$

10) Choice B is correct

The perimeter of the trapezoid is $46\ cm$.

Therefore, the missing side (height) is $= 46 - 8 - 12 - 6 = 20$

Area of a trapezoid: $A = \frac{1}{2}h\ (b_1 + b_2) = \frac{1}{2}(20)(6+8) = 140$

11) Choice A is correct

Let x be the number of new shoes the team can purchase. Therefore, the team can purchase $120x$. The team had \$20,000 and spent \$15,000. Now the team can spend on new shoes \$5,000 at most. Now, write the inequality: $120x + 15{,}000 \leq 20{,}000$

12) Choice B is correct

The probability of choosing a Hearts is $\frac{13}{52} = \frac{1}{4}$

13) Choice D is correct

First, find the sum of five numbers. $average = \frac{sum\ of\ terms}{number\ of\ terms} \Rightarrow 30 = \frac{\text{sum of 5 numbers}}{5} \Rightarrow$ $sum\ of\ 5\ numbers = 30 \times 5 = 150$

The sum of 5 numbers is 150. If a sixth number that is greater than 42 is added to these numbers, then the sum of 6 numbers must be greater than 192.

$150 + 42 = 192.$ If the number was 42, then the average of the numbers is:

$$average = \frac{sum\ of\ terms}{number\ of\ terms} = \frac{192}{6} = 32$$

Since the number is bigger than 42. Then, the average of six numbers must be greater than 32. Choice D is greater than 32.

14) Choice B is correct

Let L be the length of the rectangular and W be the with of the rectangular. Then, $L = 4W + 3$, The perimeter of the rectangle is 36 meters. Therefore:

$2L + 2W = 36,\qquad L + W = 18$

Replace the value of L from the first equation into the second equation and solve for W:

$(4W + 3) + W = 18 \rightarrow 5W + 3 = 18 \rightarrow 5W = 15 \rightarrow W = 3$

The width of the rectangle is 3 meters and its length is: $L = 4W + 3 = 4(3) + 3 = 15$

The area of the rectangle is: $length \times width = 3 \times 15 = 45$

15) Choice C is correct

The ratio of boy to girls is $3: 8$. Therefore, there are 3 boys out of 11 students. To find the answer, first divide the total number of students by 11, then multiply the result by 3.

$66 \div 11 = 6 \Rightarrow 3 \times 6 = 18$

There are 18 boys and 48 $(66 - 18)$ girls. So, 30 more boys should be enrolled to make the ratio $1: 1$

16) Choice D is correct

Isolate and solve for x: $\frac{2}{3}x + \frac{1}{6} = \frac{1}{2} \Rightarrow \frac{2}{3}x = \frac{1}{2} - \frac{1}{6} = \frac{1}{3} \Rightarrow \frac{2}{3}x = \frac{1}{3}$

Multiply both sides by the reciprocal of the coefficient of x.

$(\frac{3}{2})\frac{2}{3}x = \frac{1}{3}(\frac{3}{2}) \Rightarrow x = \frac{3}{6} = \frac{1}{2}$

17) Choice D is correct

A. $= 4 \rightarrow$ The perimeter of the figure is: $3 + 5 + 3 + 4 + 4 = 19 \neq 33$
B. $x = 6 \rightarrow$ The perimeter of the figure is: $3 + 5 + 3 + 6 + 6 = 23 \neq 33$
C. $x = 8 \rightarrow$ The perimeter of the figure is: $3 + 5 + 3 + 8 + 8 = 27 \neq 33$
D. $x = 11 \rightarrow$ The perimeter of the figure is: $3 + 5 + 3 + 11 + 11 = 33 = 33$

18) Choice D is correct

Use Pythagorean Theorem: $a^2 + b^2 = c^2 \rightarrow 4^2 + 3^2 = c^2 \Rightarrow 25 = c^2 \Rightarrow c = 5$

19) Choice B is correct

Plug in 140 for F and then solve for C.

$C = \frac{5}{9}(F - 32) \Rightarrow C = \frac{5}{9}(140-32) \Rightarrow C = \frac{5}{9}(108) = 60$

20) Choice D is correct

Simplify. $7x^2y^3(2x^2y)^3 = 7x^2y^3(8x^6y^3) = 56x^8y^6$

Accuplacer Mathematics Practice Test 2

Advanced Algebra and Functions

1) Choice D is correct

Since $N = 7$, substitute 7 for N in the equation $\frac{x-3}{5} = N$, which gives $\frac{x-3}{5} = 7$. Multiplying both sides of $\frac{x-3}{5} = 7$ by 5 gives $x - 3 = 35$ and then adding 3 to both sides of $x - 3 = 35$ then, $x = 38$.

2) Choice C is correct

$b^{\frac{m}{n}} = \sqrt[n]{b^m}$ For any positive integers m and n. Thus, $b^{\frac{4}{5}} = \sqrt[5]{b^4}$.

3) Choice A is correct

$\frac{x}{2+3} = \frac{y}{10-7} \rightarrow \frac{x}{5} = \frac{y}{3} \rightarrow 5y = 3x \rightarrow y = \frac{3}{5}x$

4) Choice C is correct

To rewrite $\frac{2+3i}{5-2i}$ in the standard form $a + bi$, multiply the numerator and denominator of $\frac{2+3i}{5-2i}$ by the conjugate, $5 + 2i$. This gives $\left(\frac{2+3i}{5-2i}\right)\left(\frac{5+2i}{5+2i}\right) = \frac{10+4i+15i+6i^2}{5^2-(2i)^2}$. Since $i^2 = -1$, this last fraction can be rewritten as $\frac{10+4i+15i+6(-1)}{25-4(-1)} = \frac{4+19i}{29}$.

5) Choice C is correct

First find the value of b, and then find $f(4)$. Since $f(2) = 35$, substuting 2 for x and 35 for $f(x)$ gives $35 = b(2)^2 + 15 = 4b + 15$. Solving this equation gives $b = 5$. Thus

$f(x) = 5x^2 + 15$, $\quad f(4) = 5(4)^2 + 15 \rightarrow f(4) = 80 + 15$, $\quad f(4) = 95$

6) Choice D is correct

Multiplying each side of $-3x - y = 6$ by 2 gives $-6x - 2y = 12$. Adding each side of $-6x - 2y = 12$ to the corresponding side of $6x + 4y = 10$ gives $2y = 22$ or $y = 11$. Finally, substituting 11 for y in $6x + 4y = 10$ gives $6x + 4(11) = 10$ or $x = -\frac{17}{3}$.

7) Choice A is correct

Identify the input value. Since the function is in the form $f(x)$ and the question asks to calculate $f(3)$, the input value is four. $f(3) \rightarrow x = 3$ Using the function, input the desired x value.

Now substitute 4 in for every x in the function. $f(x) = 3x^2 - 4$, $\quad f(3) = 3(3)^2 - 4$,

$f(3) = 27 - 4$, $\quad f(3) = 23$

8) Choice D is correct

The problem asks for the sum of the roots of the quadratic equation $2n^2 + 16n + 24 = 0$. Dividing each side of the equation by 2 gives $n^2 + 8n + 12 = 0$. If the roots of

$n^2 + 8n + 12 = 0$ are n_1 and n_2, then the equation can be factored as

$n^2 + 8n + 12 = (n - n_1)(n - n_2) = 0$. Looking at the coefficient of n on each side of

$n^2 + 8n + 12 = (n + 6)(n + 2)$ gives $n = -6$ or $n = -2$, $\quad$ then, $-6 + (-2) = -8$

9) Choice D is correct

To perform the division $\frac{3+2i}{5+i}$, multiply the numerator and denominator of $\frac{3+2i}{5+1i}$ by the conjugate of the denominator, $5 - i$. This gives $\frac{(3+2i)(5-i)}{(5+1i)(5-i)} = \frac{15-3i+10i-2i^2}{5^2-i^2}$. Since $i^2 = -1$, this can be simplified to $\frac{15-3i+10i+2}{25+1} = \frac{17+7i}{26}$

10) Choice D is correct

The x-intercepts of the parabola represented by $y = x^2 - 8x + 12$ in the xy-plane are the values of x for which y is equal to 0. The factored form of the equation, $y = (x - 2)(x - 6)$, shows that y equals 0 if and only if $x = 2$ or $x = 6$. Thus, the factored form $y = (x - 2)(x - 6)$, displays the x-intercepts of the parabola as the constants 2 and 6.

11) Choice C is correct

If $x - a$ is a factor of $g(x)$, then $g(a)$ must equal 0. Based on the table $g(1) = 0$. Therefore, $x - 1$ must be a factor of $g(x)$.

12) Choice C is correct

To solve this problem first solve the equation for c. $\frac{c}{b} = 2$

Multiply by b on both sides. Then: $b \times \frac{c}{b} = 2 \times b \rightarrow c = 2b$. Now to calculate $\frac{6b}{c}$, substitute the value for c into the denominator and simplify. $\frac{6b}{c} = \frac{6b}{2b} = \frac{6}{2} = 3$

13) Choice B is correct

Simplify the numerator: $\frac{x+(5x)^2+(3x)^3}{x} = \frac{x+5^2x^2+3^3x^3}{x} = \frac{x+25x^2+27x^3}{x}$

Pull an x out of each term in the numerator. $\frac{x(1+25x+27x^2)}{x}$

The x in the numerator and the x in the denominator cancel:

$1 + 25x + 27x^2 = 27x^2 + 25x + 1$

14) Choice B is correct

The equation $\frac{a-b}{b} = \frac{10}{11}$ can be rewritten as $\frac{a}{b} - \frac{b}{b} = \frac{10}{11}$, from which it follows that $\frac{a}{b} - 1 = \frac{10}{11}$, or $\frac{a}{b} = \frac{10}{11} + 1 = \frac{21}{11}$.

15) Choice A is correct

First write the equation in slope intercept form. Add $2x$ to both sides to get $6y = 2x + 30$. Now divide both sides by 6 to get $y = \frac{1}{3}x + 5$. The slope of this line is $\frac{1}{3}$, so any line that also has a slope of $\frac{1}{3}$ would be parallel to it. Only choice A has a slope of $\frac{1}{3}$.

16) Choice D is correct

To find the average of three numbers even if they're algebraic expressions, add them up and divide by 3. Thus, the average equals: $\frac{(4x+2)+(-6x-5)+(8x+2)}{3} = \frac{6x-1}{3} = 2x - \frac{1}{3}$

17) Choice C is correct

Let l and w be the length and width, respectively, of the original rectangle. The area of the original rectangle is $A = lw$. The rectangle is altered by increasing its length by 20 percent and decreasing its width by s percent; thus, the length of the altered rectangle is $1.2l$, and the width of the altered rectangle is $\left(1 - \frac{s}{100}\right)w$.

The alterations decrease the area by 4 percent, so the area of the altered rectangle is $(1 - 0.04)\,A = 0.96A$. The altered rectangle is the product of its length and width, therefore $0.96A = (1.2l)(1 - \frac{s}{100})w$, Since $A = lw$, this equation can be rewritten as $0.96A = (1.2)\left(1 - \frac{s}{100}\right)lw = (1.2)(1 - \frac{s}{100})A$, from which it follows that $0.96 = (1.2)\left(1 - \frac{s}{100}\right)$, divide both sides of the equation by 1.2. Then: $0.8 = 1 - \frac{s}{100}$

Therefore, $\frac{s}{100} = 0.2$ and therefore the value of s is 20.

18) Choice B is correct

Because each children ticket costs \$3 and each adult ticket costs \$4, the total amount, in dollars, that John spends on x student tickets and 2 adult ticket is $3(x) + 4(2)$. Because John spends at least \$10 but no more than \$15 on the tickets, you can write the compound inequality $3x + 8 \geq 10$ and $3x + 8 \leq 15$. Subtracting 8 from each side of both inequalities and then dividing each side of both inequalities by 3 gives $x \geq 0.66$ and $x \leq 2.3$. Thus, the value of x must be an integer that is both greater than or equal to 0.66 and less than or equal to 2.3. Therefore, $x = 1$ or $x = 2$. Either 1 or 2 may be gridded as the correct answer.

19) Choice B is correct

The sum of all angles in a quadrilateral is 360 degrees. Let x be the smallest angle in the quadrilateral. Then the angles are: $x, 4x, 6x, 7x$, $x + 4x + 6x + 7x = 360 \rightarrow 18x = 360 \rightarrow x = 20$, The angles in the quadrilateral are: $20°, 80°, 120°$, and $140°$

20) Choice D is correct

Rate of change (growth or x) is 9 per week. $45 \div 5 = 9$

Since the plant grows at a linear rate, then the relationship between the height (y) of the plant and number of weeks of growth (x) can be written as: $y(x) = 9x$

"Effortless Math" Publications

Effortless Math authors' team strives to prepare and publish the best quality ACCUPLACER Mathematics learning resources to make learning Math easier for all. We hope that our publications help you learn Math in an effective way and prepare for the ACCUPLACER test.

We all in Effortless Math wish you good luck and successful studies!

Effortless Math Authors

Made in the USA
Coppell, TX
21 April 2024